The Practical Handbook of

PLUMBING AND HEATING

By Richard Day

Fawcett Publications, Inc.
One Astor Plaza
New York, New York 10036

FRANK BOWERS: *Editor-in-Chief*

SILVIO LEMBO: *Creative Director*

HAROLD E. PRICE: *Associate Director* • HERB JONAS: *Assistant Director*

JOSEPH C. PENELL: *Marketing Director*

RAY GILL: *Editor*

LUCILLE DE SANTIS: *Production Editor*
JULIA BETZ, LORETTA ANAGNOST, *Assistants*

Editorial Staff: DAN BLUE, ELLENE SAUNDERS

Art Staff: MIKE GAYNOR, ALEX SANTIAGO, JOHN SELVAGGIO,
JOHN CERVASIO, JACK LA CERRA, MIKE MONTE

How-To Art by Henry Clark
Cover Color by the Author

Printed in U.S.A. by
FAWCETT PRINTING CORPORATION
Rockville, Maryland

SECOND PRINTING

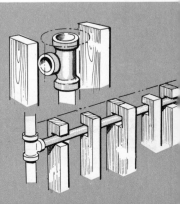

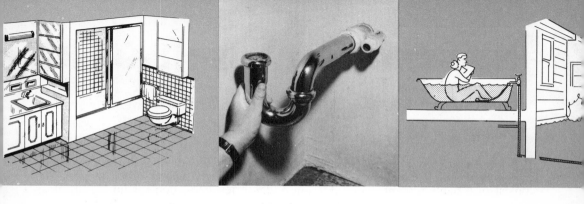

CONTENTS

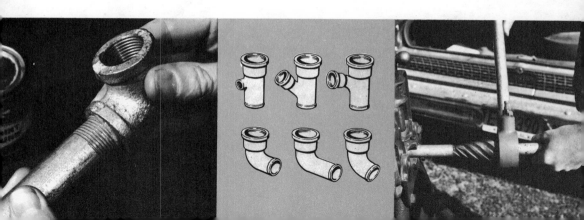

INSIDE STORY ON HEATING

Year-round comfort is the purpose of the modern home heating system

Home heating is a correct combination of many components. These consist of a heating plant, a distribution system, controls, safety devices and accessories. The heating plant — a furnace or boiler, usually — provides the energy to warm a house. The distribution system divides and circulates the heat. Controls regulate heat just as the residents want it. Safety devices prevent overheating and guard against other dangers. Accessories cool, control humidity, clean the air and do other interesting things.

The whole thing adds up to total comfort. That's what good heating is all about. If your present heating system doesn't measure up to total comfort, it may need modernizing.

Solar house makes use of low-angled rays of the sun that come through large double-glazed windows on the south wall of a house. Sunshine blankets the floors with warmth.

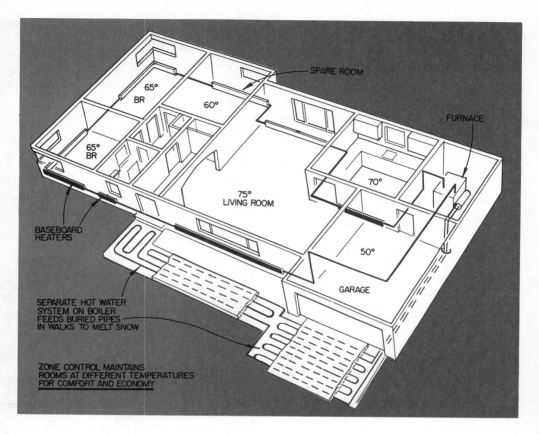

SPARE ROOM

65° BR

60°

FURNACE

65° BR

75° LIVING ROOM

70°

BASEBOARD HEATERS

50°

SEPARATE HOT WATER SYSTEM ON BOILER FEEDS BURIED PIPES IN WALKS TO MELT SNOW

GARAGE

ZONE CONTROL MAINTAINS ROOMS AT DIFFERENT TEMPERATURES FOR COMFORT AND ECONOMY

Heat is one form of energy. Energy is the power to do work; heat energy added to ice melts it. Its power to do work changes the ice from a solid to a liquid. Heat energy added to air doesn't change the physical form of the air at all. Instead it raises the air's temperature. For home heating purposes, heat is measured in two ways: temperature and quantity. The Fahrenheit degree is the unit of temperature measurement; the British thermal unit (BTU) is the unit of heat quantity.

To picture the difference between temperature and quantity of heat, visualize water in a pan on the stove. If its temperature is 68 degrees and you turn on the stove and heat it up to 78 degrees, all this tells you is how hot, not how much heat. To know how much heat was required to raise the temperature of that water 10 degrees, you need to know how much water there was. If there was 1

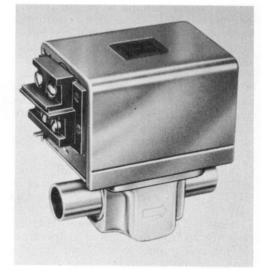

Compact zone-control valve for hydronic and steam heating lines closes off flow of heat when the thermostat is satisfied.

5

quart (2 lbs.) of water, then you can figure the quantity of heat in BTU's. A BTU is the amount of heat needed to raise the temperature of one pound of water one degree Fahrenheit. It's about the same heat as you get from burning a wooden kitchen match. In the example, two pounds of water had its temperature raised 10 degrees. This took 2 times 10, or 20 BTU's of heat.

In a heating system the burning of fuel releases BTU's which are used to heat the air in your home. The most popular fuel is gas. There are nearly 30 million customers for gas heat. Next in line is oil heat with some 10 million customers. Electric heating is coming on fast. Past the 3 million mark, electric heat is expected to have 19 million users by 1980. With electric heat, BTU's are given off as the electricity is converted to heat through tiny resistance wires. Homes beyond gas mains can still heat with gas — liquefied petroleum gas. Called L-P gas for short, it is delivered by truck. Coal as a home heating fuel gets a cool reception by most homeowners.

HEAT LOSS

In practical heating work the amount of heat that escapes from a house through its walls, floors, and ceilings is figured in BTU's per hour. Called *heat loss*, this can be calculated for any indoor-outdoor temperature difference. The coldest temperature it's likely to get during a severe winter is chosen as a *design temperature*. Design temperature varies from area to area. Heat loss is figured at design temperature outdoors and 75 degrees indoors.

Since the BTU is such a small amount of heat, it takes many thousands of BTU's per hour to heat a home on a cold day.

Heat is like a liquid, it flows. Heat always flows from the warmer temperature to the colder one. Heat flows out of a house in cold weather, into it in hot weather. The greater the temperature difference, the greater the heat flow.

Heat flow can take place in three

HEAT IS DRAWN FROM BODY TO COLD WALL   WITH HEAT ON WALL, NO BODY HEAT IS LOST

Your body radiates heat to a cold wall. When walls are properly warmed, body heat can be kept at a pleasant, proper level.

Radiated heat reaches and warms all surfaces and objects in room. Convected heat circulates through a room as warmed air.

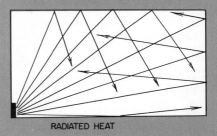

RADIATED HEAT

CONVECTED HEAT

ways: conduction, convection and radiation. Conducted heat travels through a substance, for example, through the wall of a house. Convected heat moves in a fluid. Air is a fluid. So is water. Both are used to move heat around in a heating system. Radiated heat travels through space. The sun's heat reaches us by radiation. Radiated heat doesn't warm the air it travels through, only the object it lights on. All objects radiate heat. The higher their temperature the more heat they radiate.

A heating system isn't supposed to be something to warm people occupying a building in cold weather. People generate their own heat. What a heating system must do is help the body regulate the rate at which it gets rid of its excess heat.

The principal things that affect comfort are air temperature, movement, humidity and radiation of heat from the body. Since comfort depends partly on heat radiated from the body, a room with a large expanse of windows seems cooler than a small-windowed room at the same temperature. Heat radiates from your body into and through the cold window panes. Less heat is radiated to a *warmer* wall. A room without drafts seems warmer, too, than one in which icy air brushes against you.

A heating system that can do the following is sure to make a comfortable home: maintain a 75-degree temperature in every room with no more than 3 degrees variation room-to-room; keep a relative humidity of up to 30 percent; prevent cold drafts from large windows; and hold the floor-to-ceiling temperature differential to a minimum.

Five kinds of heat can be used in the house heating system: warm air, hot water, steam, radiant electric and unit heaters. Though steam heating was once popular, it's rarely used in houses today. Hot water heat has replaced it.

WARM AIR HEAT

With warm air heating, the heating plant is called a furnace. A plenum chamber, usually on top of the furnace, collects heat. On modern forced-air systems a blower forces heated air out of the furnace plenum. Old-style systems work by gravity. Forced-air furnaces may be up-flow, down-flow or horizontal-flow, depending which best suits the house. Ducts, 4-inches in diameter and larger, connect to the plenum and channel warm air around to the rooms. In the rooms, wall diffusers or floor registers direct the warm air into the room to best counteract cold drafts. Return air is picked up at one or more central locations and ducted back to the furnace, runs through a filter and is then reheated and recirculated.

Warm-air heating systems are designed and installed according to the manuals of instruction issued by the National Warm Air Heating & Air Conditioning Association. The systems are classified by arrangement of ductwork and heating outlets. The simplest uses pipes running from the hot-air plenum directly to the room outlets. There may be more than one outlet per room arranged around the perimeter of the house. This simple pipe system is especially suited to square or rectangular floor plan homes with basements.

Another type that's popular for a slab-on-ground basementless house is the perimeter loop system. A looping duct embedded in the concrete slab is fed by radial ducts, also embedded in the slab. A down-flow furnace is used with this system. Cold air enters at the top of the furnace and heated air is blown out the bottom. Return air ducts in the perimeter loop system are usually run above the ceiling.

A crawlspace plenum system can be used on one-story houses built over a crawlspace. The furnace and registers have no ducts connecting them. Warm air is blown directly into the crawlspace through short ducts that aim it toward the outside walls. Registers placed around the perimeter of the house open into the crawlspace and let heated air up into the rooms. Warm crawlspace air also warms the floors by conduction. Such a crawlspace should be well insulated and the ground covered with a vapor barrier.

Still another type of warm-air heating is called the extended plenum system. A

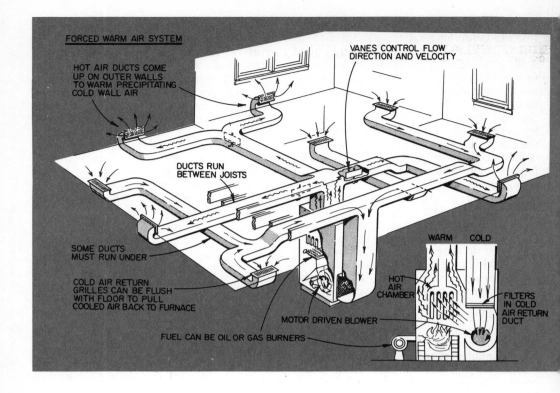

HOT AIR DUCTS COME UP ON OUTER WALLS TO WARM PRECIPITATING COLD WALL AIR

VANES CONTROL FLOW DIRECTION AND VELOCITY

DUCTS RUN BETWEEN JOISTS

SOME DUCTS MUST RUN UNDER

COLD AIR RETURN GRILLES CAN BE FLUSH WITH FLOOR TO PULL COOLED AIR BACK TO FURNACE

FUEL CAN BE OIL OR GAS BURNERS

MOTOR DRIVEN BLOWER

WARM   COLD

HOT AIR CHAMBER

FILTERS IN COLD AIR RETURN DUCT

large duct, or ducts, take off from the hot-air plenum and run for most of the length of the house. Smaller ducts take off from these to serve room outlets. The extended-plenum system is adaptable to long, rambling houses with basement or attic furnaces.

Advantages claimed for warm-air heat are filtration, humidification and easy adaptability to air conditioning. By adding an outdoor air intake, a forced-air heating system will deliver freshened air to all rooms. Constant mixing of the air is said to minimize temperature differences from room to room and between floor and ceiling.

HOT-WATER HEATING

Hot-water heat, also called hydronic heat, uses a boiler to heat water that's circulated through finger-size pipes to all rooms of the house. A pump provides the push. In each room a radiator releases heat from the water into the room by both radiation and convection. Radiators are usually installed in baseboard units or convectors that prevent anyone from coming in contact with a hot radiator surface. Yet they allow air to circulate over them. Forced-hot-water systems are designed and installed according to criteria of the Institute of Boiler and Radiator Manufacturers.

Hot-water heating has three variations, each named for the way the distribution pipes are arranged. The *series loop* system is the least expensive. In it, hot water travels through the first radiator, then the second, then the third and so on until it gets back to the boiler for reheating. No radiator valves can be used. Shutting one radiator off would stop all circulation. There is little control of temperatures, room to room, unless more than one series loop is used.

The one-pipe hydronic system is better than the series loop. It will take care

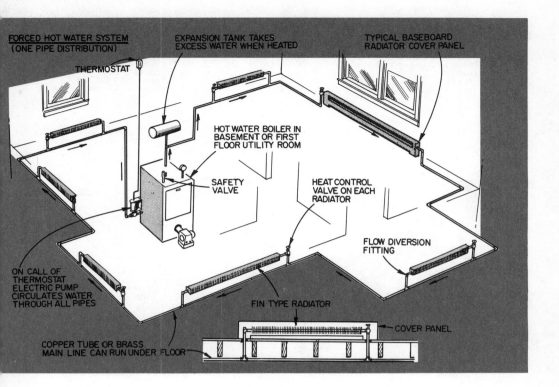

FORCED HOT WATER SYSTEM
(ONE PIPE DISTRIBUTION)

THERMOSTAT

EXPANSION TANK TAKES
EXCESS WATER WHEN HEATED

TYPICAL BASEBOARD
RADIATOR COVER PANEL

HOT WATER BOILER IN
BASEMENT OR FIRST
FLOOR UTILITY ROOM

SAFETY
VALVE

HEAT CONTROL
VALVE ON EACH
RADIATOR

FLOW DIVERSION
FITTING

ON CALL OF
THERMOSTAT
ELECTRIC PUMP
CIRCULATES WATER
THROUGH ALL PIPES

FIN TYPE RADIATOR

COVER PANEL

COPPER TUBE OR BRASS
MAIN LINE CAN RUN UNDER FLOOR

of any but the largest houses. A single supply pipe loops around to all radiators and returns to the boiler. Each end of every radiator is connected to the supply loop by a *flow-diversion fitting*. These fittings are like pickpockets. Each one grabs off a little of the circulating hot water, runs it through the radiator and back to the supply loop again. The rest of the water flows past that radiator. Adjustment is provided for controlling the amount of hot water each radiator receives.

The two-pipe hydronic system is used in large, expensive houses. It's like the one-pipe system but provides separate supply and return piping systems.

Forced-hot-water heating advocates say it achieves a balance between radiated and convected heat, making you feel comfortably surrounded by a "ring of radiant warmth." Sounds nice. Hot-water heat is claimed to be better, since dust, odors and germs are not circulated. Hy-dronic heating is simple to install. Zone control is easy with it too, by supplying different amounts of heat to different parts of the house.

Steam heat's one-pipe and two-pipe systems are similar to hot-water systems except steam circulates itself through the system. No pump is used. The return trip to the boiler is by gravity with steam in its liquid state.

RADIANT HEATING

Radiant heating, whether electric or hot-water, is said to be the ultimate in comfort, like being in the sun on a pleasant spring day. Moreover, there's nothing of the heating system to be seen — no radiators, no registers. Just pipes hidden in the walls, floors and ceilings.

The warmed surfaces give a combination of conducted and radiated heat to the room. Rugs and other insulating floor coverings over a radiant-heated floor

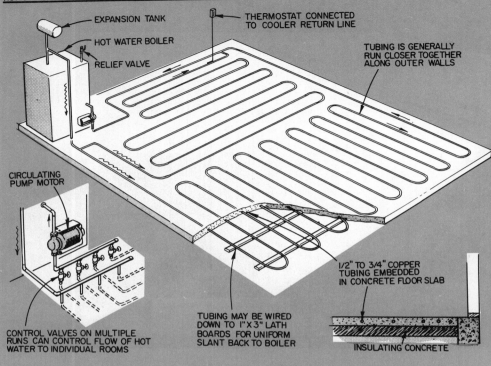

EXPANSION TANK

THERMOSTAT CONNECTED TO COOLER RETURN LINE

HOT WATER BOILER

RELIEF VALVE

TUBING IS GENERALLY RUN CLOSER TOGETHER ALONG OUTER WALLS

CIRCULATING PUMP MOTOR

CONTROL VALVES ON MULTIPLE RUNS CAN CONTROL FLOW OF HOT WATER TO INDIVIDUAL ROOMS

TUBING MAY BE WIRED DOWN TO 1"X 3" LATH BOARDS FOR UNIFORM SLANT BACK TO BOILER

1/2" TO 3/4" COPPER TUBING EMBEDDED IN CONCRETE FLOOR SLAB

INSULATING CONCRETE

RADIANT ELECTRIC CEILING

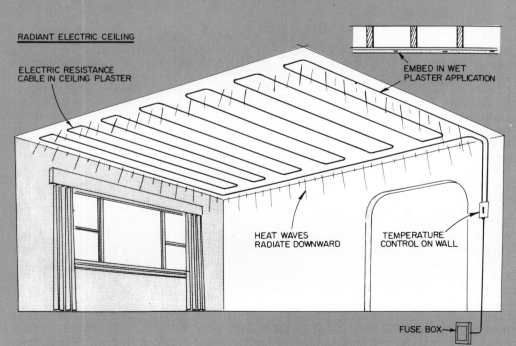

ELECTRIC RESISTANCE CABLE IN CEILING PLASTER

EMBED IN WET PLASTER APPLICATION

HEAT WAVES RADIATE DOWNWARD

TEMPERATURE CONTROL ON WALL

FUSE BOX

10

create balancing problems by getting in the way of heat transmission. Floor surfaces in a radiant system are usually at less than 80 degrees. Ceilings normally don't get hotter than 120 degrees.

Temperature regulation can be tricky because radiant hot-water heating is slow to respond to changes in heating needs. It takes time to heat up and cool off.

ELECTRIC HEATING

Electricity can be used as the "fuel" in hot-water or forced-air heating systems. But it can also be routed to resistance wires or panels in rooms, and be used directly as a type of heating system.

Electric utilities are competing hard for the home heating market. Because of the high fuel cost, electrically heated homes are usually insulated and weather-stripped to the hilt. The idea is to keep heat loss to a minimum. This adds to the initial cost of electric heating.

Electric heating is great. But it tends to be doggone expensive, at present, anyway. Those living in TVA areas and the Pacific Northwest have an advantage. Rates there are already low.

With an electric rate of 1½ cents per kilowatt hour or less this type of heat may be worth thinking of. You'll usually need a 200-ampere electrical service entrance and 220-volt service. Insulation will be needed on the order of 6 inches in the ceiling, 4 inches in the walls and 2 inches in the floors. Many electric utilities offer all kinds of inducements to get you started heating electrically. Check with your utility. What's more, look into things such as cost of adding insulation to an existing house, costs of installing the various systems and cost of maintenance (some types of electric heat have no moving parts). Also, if you'll want air conditioning, you may need a system of ducts anyway. Don't count on saving the expense of installing them.

Ceiling Cable—One popular method of electric heating is with resistance cables embedded in the ceiling to furnish radiant electric heat. These are used with plastered or plasterboard ceilings. The

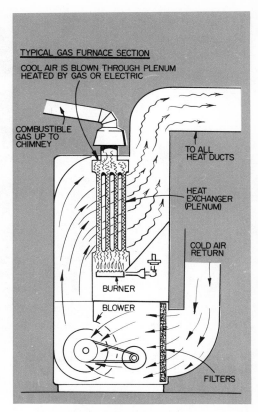

TYPICAL GAS FURNACE SECTION
COOL AIR IS BLOWN THROUGH PLENUM HEATED BY GAS OR ELECTRIC
COMBUSTIBLE GAS UP TO CHIMNEY
TO ALL HEAT DUCTS
HEAT EXCHANGER (PLENUM)
COLD AIR RETURN
BURNER
BLOWER
FILTERS

Above, components of furnace that create heat, transfer it to air, circulate it to rooms. Combustion products go up chimney.

Electric boiler is so compact it fits on wall. Piping is same as with any boiler.

The heater's resistance wires fit into heating duct. Wiring connects to house system.

wires are fastened to the ceiling before installing the finished wall surfacing.

Special plasterboard panels containing a conductive film are installed with the dry-wall radiant heating method. It's really catching on. U.S. Gypsum makes a panel called "Thermalux Electric Heating Panel." The initial cost for heating with this system is said to be between that of hot-water and forced-air. You can probably install the panels yourself if you get full directions on how to do it.

In another system a vinyl sheet containing copper mesh conductor in nylon netting is adhered to the ceiling as in wallpapering and wired into the system.

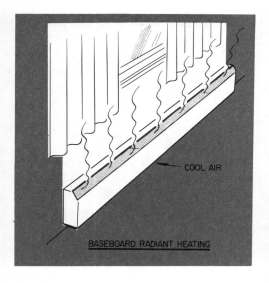

COOL AIR

BASEBOARD RADIANT HEATING

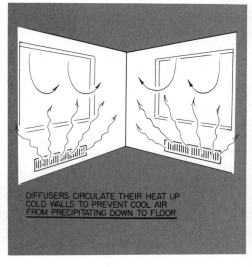

DIFFUSERS CIRCULATE THEIR HEAT UP COLD WALLS TO PREVENT COOL AIR FROM PRECIPITATING DOWN TO FLOOR

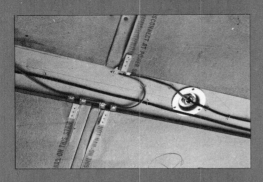

Plasterboard heating panel is the newest thing in electrical heating. Panels are cut to fit, nailed to ceiling and wired.

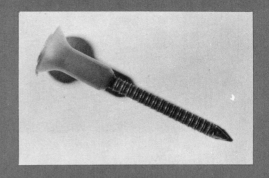

Use nails insulated with plastic shoulders to keep metal from coming in contact with the resistance-heating layer in each panel.

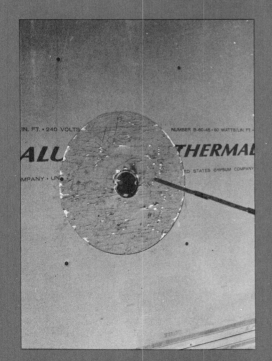

Resistance layer in panel's surface is cut back around fixture boxes and other metal projecting through it to prevent burn-out.

You need insulation above every ceiling with this type of heat to make heat travel down into room, not up to floor above.

Apply adhesive to back of ½-inch plaster-board to ready it for stick-on insulation.

It can be painted over. With thermostat control, an electrically wired-up room becomes a giant-sized radiant heater. No duct-work is needed. Comfort is supposed to be outstanding. Insulation is re-quired between all floors and on all walls and ceilings.

Heating Units — Next most popular are electric baseboard heating units. These are mounted near the floor, usually along outside walls of rooms and especially under windows. Air flows through them and up the wall, keeping cold drafts from settling to the floor. They're controlled by a thermostat, one per room, if desired. Air circulation is by convection. The benefits of perimeter heating can be had in this way. You can gang up as many baseboard units as the house electrical system will stand. A whole house can be heated this way if its electrical system has enough zap. You can also get radiant heated panels that mount below a window or in a too-cool room to provide supplementary heat.

Electric Central Heating — An electric furnace is much like a gas or oil furnace with a blower and ducts to distribute the heat provided by resistance wires in the furnace. Duct heaters are mini-furnaces that fit right in a large duct. Electric boil-ers for hydronic heating use electricity instead of gas or oil to heat water in the boiler. From there, the system is the same as any other hot-water heating sys-tem.

Heat Pump — The heat pump can heat or cool a house, depending on which di-rection it moves the heat. During the heating season, it wrings heat out of the

Finished plasterboard ceiling panels are put on and then nailed on to the heating panels.

Heating a home addition is easy if you use electric valance heaters on outside walls.

air outside your home, raises its temperature by concentrating it and brings the heat inside. During the cooling season the heat pump squeezes heat out of room air, concentrates it and pumps it outside. Heat pumps are most common in milder climates where there is enough heat in the winter air. Manufacturers claim that refinements make heat pumps practical even in colder climates. Some pumps use water instead of outdoor air as the medium.

HEAT FROM THE SUN

Solar heating has been around for some time, but isn't yet practical for most heating needs. However, since the sun's heat is free, this method has appeal. The simplest kind of solar heat makes use of large double-glazed windows on the south wall of a house. With these the sun

can shine into the house, blanketing its floor with warmth. In a house with floor-to-ceiling windows on most south rooms, the heating plant may shut off in mid-morning and stay off until after dark even on the coldest sunny days. Some solar heat is available on cloudy days. Since this heat is not stored, a full-fledged heating system is required to heat the house at night.

A drawback, the large windows that let heat in during the day, let heat out at night. They should be provided with insulating draw-draperies.

A generous roof overhang is needed in a solar house to provide shade for eliminating the sun's rays altogether in summer months when the sun is high in the sky.

Another type of solar heat uses heat-collecting panels on the roof to heat water, which is then stored in a large insu-

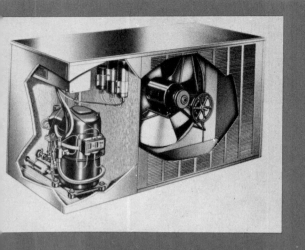

Wrap-around filter media placed in furnace removes large dust particles from air.

Heat pump's mechanism is reversible to let heat be taken from or put into the house.

The heated air is circulated through furnace and ductwork by squirrel-cage blower.

Electric resistance furnace works like a hair dryer with the air being blown over glowing wires of heating elements, right.

Photos from Lennox Industries

lated tank. The hot water is circulated as needed to radiators around the house, the same as in any hydronic heating system. A small standby heating unit provides heat when solar energy won't do the job by itself. The cost of heat-collecting panels currently limits this type of solar heating to warm, sunny climates. You can't get a straight answer out of anyone as to which kind of heating is best. Each one warms up to his own type of system and runs cool on the others. The truth is that any of the modern heating systems can be fully comfortable,

clean, quiet and completely satisfactory if properly designed, installed and adjusted. Get good advice on design from experts. The installation and adjustment you can do yourself.

UNIT HEATERS

Add-on heating for supplying additional heat to hard-to-warm rooms or house additions may be done with unit heaters. These can be gas, oil or electric. Don't expect them to heat a whole house uniformly. And don't look for the heating comfort you'd get from a well de-

Floor furnace—This fits into a hole cut in the floor with a grating over it. Heat comes up through the grating and warms the room. Cold air settles back through the outside of the same grating. Thermostat control is available. Floor furnaces are made in oil or gas. An outside vent or chimney connection is necessary. A floor furnace is capable of heating more than one room.

In-the-wall heater—Oil, gas or electric, the wall heater is normally installed between studs in an outside wall. Air comes in at the bottom, is heated and blown out at the top. Venting with gas is through the wall to a combination intake-exhaust fitting. This does away with the need for a vent stack to the roof. Most wall heaters have circulating fans and thermostats. They'll heat more than one room. Both convection and radiant heat are produced.

Fireplace—A fireplace is a born loser's method of heating a house. Most of the heat goes up the chimney. The newer circulating fireplaces built around metal units reclaim more of the heat. Even they won't supply more than emergency heat in cold weather. Enjoy a fireplace, but don't look to it for a fulltime heat source.

Portable heaters—These electric jobs come with or without blowers to help circulate heat. They're plugged into a 110-volt outlet and operated as needed to warm a small room, such as a bathroom, or to take the chill off a cool room while it is being used.

Gee whiz stuff—Such dreamy additions to a heating system such as air-ionizing and ultraviolet air treatment may someday be commonplace. Lennox Industries now offers an activated charcoal purifier to remove chemical and gaseous contaminants from circulated air.

Put it all together and you have modern climate control, summer and winter, day and night. Today's heating system has come a long way from the old potbelly. Who knows what the future will bring, perhaps a personal heating unit that blankets you with sunny warmth wherever you are. Then home heating would be less important. Until then, modernize your heating system and keep it working at top efficiency.

signed central heating system. The unit must match the fuel.

Unit heaters heat by both convection and radiation. The different types are: room heaters, in-the-wall heaters, fireplaces, and portable heaters.

Room Heater—Available in gas, oil or electric, some models of room heaters are equipped with thermostats. They're the modern version of the coal- or wood-burning stove. You can't cook on 'em. Gas and oil types require a flue. Some come with blowers for circulating heat across the floor and around the room.

Don't touch that dial or you'll be a thermostat-jiggler. Set the thermostat where you and your family are most comfortable and leave it. That's the modern way.

MEET YOUR HEATING CONTROLS

Controls function automatically to give you service and safety.

The modern heating system has many controls. They are of two kinds: operating controls and safety controls. Operating controls regulate the flow of fuel to the heating plant and turn it on and off. The thermostat is the most obvious one. There are others. Safety controls protect you and your family from unsafe conditions connected with the fuel or the heat. One of these is the flame-sensing device that shuts off the fuel supply if a burner doesn't light.

Your thermostat is the heating system's envoy to your house. It tells the heating system what temperature you want the house to be kept. The thermostat can only ask for heat. It's up to the

heating system to deliver. A good thermostat can control temperatures so closely no one will ever notice any variation. Sometimes more than one thermostat is used for zone-controlled heating. A thermostat is placed in each zone.

Thermostats are merely electric switches. Most of them work on 24 volts of current. In many, a temperature-sensitive bimetallic spring inside reacts to temperature changes. As it does, it opens and closes a set of electrical contacts. The contacts say "go" or "stop" to the burner or, in electric heating, the resistance elements.

A thermostat is affected only by the air that comes in contact with it. Therefore, it should be located in the most lived-in room, usually the living room. Don't put a thermostat in interior passages such as hallways. These are slowest to take notice of house temperature changes. The thermostat shouldn't be placed in rooms directly above the heating plant. These rooms aren't typical of the rest of the house. Keep a thermostat off a warm wall, behind a refrigerator or stove and away from radiators or registers. Don't locate it on an outside wall either. Heating experts recommend putting the thermostat on an inside partition about 18 inches from an outside wall, not near doors to the outside. The wall shouldn't be in the sunlight at any time during the day. Keep radio, television and lamps away from your thermostat. Their heat tends to mislead it. Air circulation around the thermostat should be good, but don't place it in a draft.

The height of the thermostat is important too. Ideally it should be 2½ to 3 feet from the floor. But if small children roam the house, locate it four feet high, out of their reach.

MODERN SYSTEMS

The most modern heating systems have both indoor and outdoor thermostats. The outdoor ones sense outside temperature changes and adjust the temperature of the heating medium — water or air — for comfort. Homes that have radiant heating especially need this dynamic thermostatic duo to give the heating system advance notice of temperature changes. This keeps the system in tune with the weather. Some stats are so sensitive that a 1/10th degree change in temperature will adjust the burner flame.

Most thermostats have small resistance heating coils below the temperature-sensing device. When the thermostat is calling for heat, its coil gives the sensing element a hotfoot, making it cycle off sooner than it otherwise would. Residual heat in the furnace or boiler brings up the room temperature to the desired point. As soon as the thermostat switches off, the heating element cuts off too, letting the temperature sensor cool

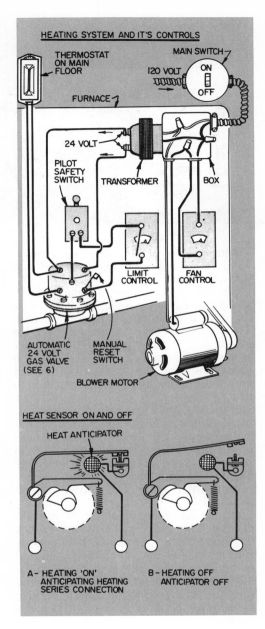

again. The effect is to make the thermostat extremely sensitive to temperature variations.

An air conditioning thermostat is like this, but it works in reverse. It has a heater that operates while the thermostat is switched *off*, rather than on.

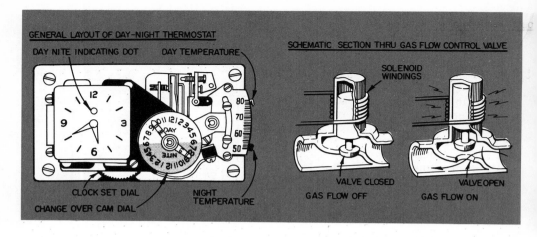

GENERAL LAYOUT OF DAY-NIGHT THERMOSTAT

DAY NITE INDICATING DOT   DAY TEMPERATURE

CLOCK SET DIAL   NIGHT TEMPERATURE

CHANGE OVER CAM DIAL

SCHEMATIC SECTION THRU GAS FLOW CONTROL VALVE

SOLENOID WINDINGS

VALVE CLOSED   VALVE OPEN
GAS FLOW OFF   GAS FLOW ON

Many types of thermostats are available. Some are combinations that will serve both heating and cooling.

Should you set your thermostat back at night, as many people do? There is a theoretical 6 percent saving to be had. The hangup is that you may not realize any saving at all by night setback. It depends on how much of a setback you give, how long the setback is in effect and how the heating system responds to it.

Experts offer these rules for night setback: Set back for sleeping comfort, not for economy. Don't set back more than 6 degrees. Any saving may be wasted in reheating your house in the morning. Skip your setback when the outdoor temperature gets really low, and on windy nights.

If you dig night setback, look into replacing your present thermostat with a day-night clock-thermostat that does the setting for you automatically. Then you won't have to get up to a cold house. Today's trend is to set the thermostat and forget it. Modern indoor climate control makes a year-round setting of 75 degrees practical.

Face it. Even if you could save 6 percent of your heating bill, would it be worth the discomfort of getting up to a cold house?

CIRCULATION CONTROLS

A blower or pump control is usually located on the furnace plenum or on the boiler water jacket. This controls circulation of heated air or water. When the burner warms the heating medium sufficiently, the control switches on the blower or pump motor. Even after the burner stopped, the motor keeps running until the temperature of the heating medium has fallen to a predetermined level. Then the control switches off its motor.

For safety, a limit control is needed on burner operation. If the temperature in the furnace or boiler should get too high, as from a clogged air filter or airlocked water pipes, the limit control will shut off the burner. The burner will stay off until all excess heat has been dissipated. Then it may come on again. The limiting operation is fully automatic. There's nothing to reset. The temperature in forced-air furnaces and open hydronic systems shouldn't exceed 200 degrees. On many systems both circulation control and limit control are in one unit.

On boilers there's also a pressure switch to cut off burner operation when the boiler reaches operating pressure. And there's a low-water cutoff float in the boiler to stop the burner if the water level drops dangerously low. An additional safety blowoff valve opens mechanically to relieve excess boiler pressure.

A solenoid control for opening the fuel valve when the thermostat signals "on" is used in gas heating. They function on

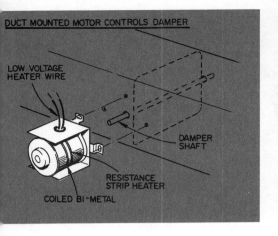

DUCT MOUNTED MOTOR CONTROLS DAMPER

LOW VOLTAGE
HEATER WIRE

DAMPER
SHAFT

RESISTANCE
STRIP HEATER

COILED BI-METAL

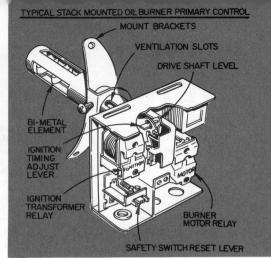

MOUNT BRACKETS

VENTILATION SLOTS

DRIVE SHAFT LEVEL

BI-METAL
ELEMENT

IGNITION
TIMING
ADJUST
LEVER

IGNITION
TRANSFORMER
RELAY

BURNER
MOTOR RELAY

SAFETY SWITCH RESET LEVER

low voltage supplied directly through the thermostat. Thermostat resistance and solenoid valve resistance should be compatible.

Automatic pilot controls on a gas burner are provided to shut off the burner's gas supply should the pilot flame snuff out. Often a thermocouple heated by the pilot flame does the trick. Some systems shut off only the main burner gas. Because L-P gas is heavier than air, its safety control must shut off both the burner and the pilot gas. Such a unit is called a 100-percent-shutoff pilot. One is required with each L-P gas appliance to prevent gas from collecting in a basement or crawlspace when the pilot goes out.

Electric heating systems are switched on with relays that isolate the low thermostat draw from the high heating element draw. Time-delay controls often are used to cycle one set of elements, then another and then another, to prevent overloading the house power lines unnecessarily.

PRIMARY CONTROL

Most oil burners have what is called a *primary control*. It's loaded with relays and other complicated devices. The primary control stays ready at all times to switch the oil burner on or off, depending on what message it receives from the thermostat. The primary control always remains subordinate to the limit controls,

however. If they say "no," the primary control can't start up the burner.

The primary control supervises starting, running and stopping of the burner. Its most important job is to tell whether there's a flame in the firebox during the initial start as well as during the running period. All burner primary controls begin a "safety" program at the start of each cycle. This shuts off the burner if a flame doesn't catch.

Primary controls can be stack-mounted; burner-mounted with a stack-mounted sensor; burner-mounted with a heat-sensitive sensor in the firebox or air tube; or burner-mounted with a light-sensitive flame detector. The light-sensitive ones are the quickest-operating.

Mechanical draft controls on all burners eliminate downdrafts that might blow out a gas pilot and prevent excessive chimney drafts from affecting the fire in an oil burner.

Controls are designed to be mostly trouble-free, and good ones are. Once in a while they need service. Call a serviceman unless you know what you're doing.

High quality forced air furnaces have Palm Beach blower controls to run blower faster or slower, depending on bonnet temperature.

HELPFUL HEATING ACCESSORIES

One unit will eventually take care
of heating and cooling the home

The goal of heating experts today is building total comfort into a home heating system. A heating plant and distribution system alone won't do it. You need some heating (and cooling) accessories.

Air conditioning is bigger than heating. The day is coming when your heating system won't be up-to-date unless it cools too. A cooling system may actually be separate from the heating system with a single central refrigeration unit outdoors, and cooling coils inside ducts to create and distribute cool air. The size of the cooling unit must be based on the cooling load of your house, just as the heating system is based on its heating load. Proper design will enable the cooling system to keep the house at 75 degrees and 50 percent relative humidity all summer long. Other temperature-humidity combinations are possible if preferred.

The size or capacity of cooling equipment is figured in "tons" or BTU's of heat-removing capacity per hour. One ton of cooling is the amount that would be done by a ton of ice melting in 24 hours. It figures out to about 12,000 BTU's each hour.

Don't get an air conditioning unit that's too far oversized for the load, or it won't run enough to control house humidity. And don't get a unit too far un-

Automatic snow-melting can be added to a hydronically heated house by simply burying pipes in the concrete and then hooking the buried pipes into the home heating system.

Better Heati-Cooling Council

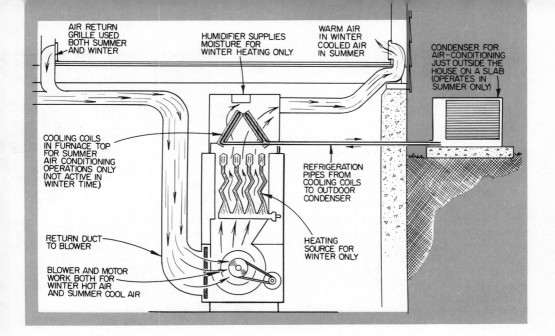

AIR RETURN GRILLE USED BOTH SUMMER AND WINTER

HUMIDIFIER SUPPLIES MOISTURE FOR WINTER HEATING ONLY

WARM AIR IN WINTER COOLED AIR IN SUMMER

CONDENSER FOR AIR-CONDITIONING JUST OUTSIDE THE HOUSE ON A SLAB (OPERATES IN SUMMER ONLY)

COOLING COILS IN FURNACE TOP FOR SUMMER AIR CONDITIONING OPERATIONS ONLY (NOT ACTIVE IN WINTER TIME)

REFRIGERATION PIPES FROM COOLING COILS TO OUTDOOR CONDENSER

RETURN DUCT TO BLOWER

HEATING SOURCE FOR WINTER ONLY

BLOWER AND MOTOR WORK BOTH FOR WINTER HOT AIR AND SUMMER COOL AIR

dersized or it won't handle the load on really hot days. An average U.S. home needs from two to four tons of cooling, depending on its size and location. Cooling capacity figures are listed on the nameplates of modern cooling units.

Air-conditioning systems can be either the through-the-window or through-the-wall kind or centrally ducted. Fuels are usually electricity, but can be gas, too.

Sometimes cooling can be added to a forced-air heating system. Other times ideal cooling requires larger ducts, a more powerful blower and cool air outlets that are placed high on the wall rather than on the floor. Don't sacrifice good cooling by shackling the air conditioner to a heat-distribution system that's not designed for cooling as well as for heating.

DEHUMIDIFICATION

Sometimes removal of moisture from the air makes life more comfortable, even without cooling. Basement rooms in most climates nearly always need dehumidification in summer. If the house isn't air conditioned, moisture-removal is accomplished with a separate unit, usually one on casters that can be rolled

where it is needed. If placed over a basement drain, the unit never needs emptying. Otherwise, collected water must be dumped regularly to avoid overflowing.

With a gauge called a *humidistat* that controls operation of the dehumidifier, you can keep your home's relative humidity at reasonable levels to prevent mildew. Most units have a humidistat built in. If not, a plug-in type can be used. The rooms being dehumidified must be kept closed. No machine can dehumidify the outdoors. One unit, if it has the capacity, will dry all the rooms open to it.

HUMIDIFICATION

The opposite of removing moisture from the air, humidification, puts water into the air. It's almost always needed during the heating season when heated air tends to be dry.

Humidity—water vapor in the air—is so important in imparting the feeling of warmth that with ample relative humidity, the temperature can be lower and still be comfortable. Tests have shown that most people are comfortable at a combination of 30 percent relative humidity and 75 degrees. For a drop of 30 percent in relative humidity, the room

23

temperature must be raised five degrees to maintain the same body comfort. Such an increase in indoor temperature would show a 10 to 15 percent increase in your fuel bill.

Outside winter air at 20 degrees and 40 percent relative humidity, when heated to 72 degrees inside your house, dries out to only 6 percent relative humidity. This is drier than the Sahara Desert. Humidity should stand at 30 percent, except in very cold weather. There aren't many houses where the relative humidity in winter can be above 30 percent without condensation problems on walls and windows. At this humidity, a temperature of 75 degrees is comfortable to most people (98 per cent of those tested). This corresponds to the temperature at which most people keep their thermostats.

Some of the needed moisture in a house is put back into the air by breathing, washing, cooking, showering, etc. But this is often only enough to keep the humidity level at 10 to 15 percent during cold weather. The rest of the moisture must be supplied artificially by a humidifier.

The small plate-type humidifiers that come with many furnaces and mount to the hot air plenum won't add enough water to be worthwhile. They lack evaporative power. You need one of the big, powerful units that literally throw water into the air. As much as 8 to 16 gallons of water a day may be needed. A good unit that can handle this job costs upward of $50. You may save that in doctor bills the first year. Colds and other respiratory infections are greatly reduced by having proper home humidity. I've personally experienced this. Didn't believe it when my doctor told me. Static shocks are done away with too.

Minerals in the water are a problem with humidifiers. They build up and eventually affect performance. The best ones are self-cleaning. Lennox makes one that flushes away mineral deposits automatically.

A humidifier is best when controlled by a humidistat. The humidifier can operate long enough to reach the desired humidity, then shut off.

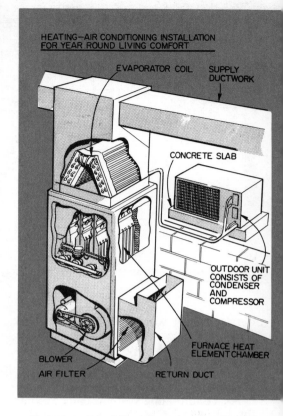

HEATING—AIR CONDITIONING INSTALLATION FOR YEAR ROUND LIVING COMFORT

EVAPORATOR COIL — SUPPLY DUCTWORK — CONCRETE SLAB — OUTDOOR UNIT CONSISTS OF CONDENSER AND COMPRESSOR — FURNACE HEAT ELEMENT CHAMBER — RETURN DUCT — AIR FILTER — BLOWER

Relative humidity to strive for is: 15 percent at 20 degrees below zero; 20 percent at 10 below; 25 percent at 0; and 30 percent at 10 above and higher.

To help the humidifier, your home should be tightly calked, weatherstripped and have a continuous vapor barrier around the walls and ceilings. If there isn't a vapor barrier, you can paint one on with an oil-base paint. Once that's on you can paint over it with any kind of paint.

ELECTRONIC AIR-CLEANING

Dust settling on furniture is like an iceberg. It warns of a huge unseen dirty air problem. A cubic foot of city air may contain more than 400 million unseen dirt particles floating around. Pollen, tobacco smoke and other irritants are there, too. Modern electronic air-cleaning can eliminate 95 percent of the air-

Air-conditioning cooling coils are placed in furnace airstream and then are piped to an outdoor condenser.

Lennox Industries

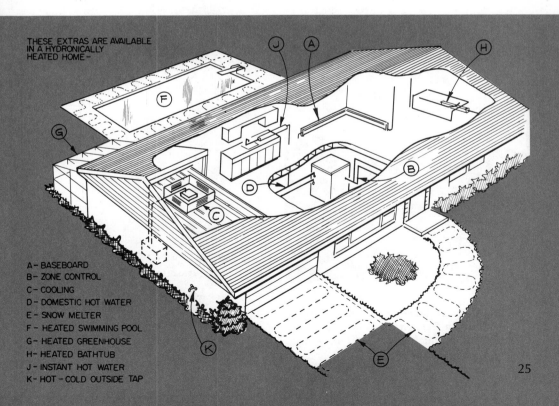

THESE EXTRAS ARE AVAILABLE IN A HYDRONICALLY HEATED HOME —

A – BASEBOARD
B – ZONE CONTROL
C – COOLING
D – DOMESTIC HOT WATER
E – SNOW MELTER
F – HEATED SWIMMING POOL
G – HEATED GREENHOUSE
H – HEATED BATHTUB
J – INSTANT HOT WATER
K – HOT – COLD OUTSIDE TAP

25

Before you fasten it tightly with sheet metal screws, level furnace humidifier on the metal hot-air plenum.

Lennox Industries

An outdoor air intake pipe on a forced-air system can freshen up your home. The pipe leads to the furnace cold-air return.

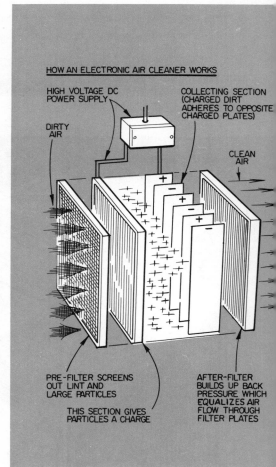

HOW AN ELECTRONIC AIR CLEANER WORKS

HIGH VOLTAGE DC POWER SUPPLY

COLLECTING SECTION (CHARGED DIRT ADHERES TO OPPOSITE CHARGED PLATES)

DIRTY AIR

CLEAN AIR

PRE-FILTER SCREENS OUT LINT AND LARGE PARTICLES

THIS SECTION GIVES PARTICLES A CHARGE

AFTER-FILTER BUILDS UP BACK PRESSURE WHICH EQUALIZES AIR FLOW THROUGH FILTER PLATES

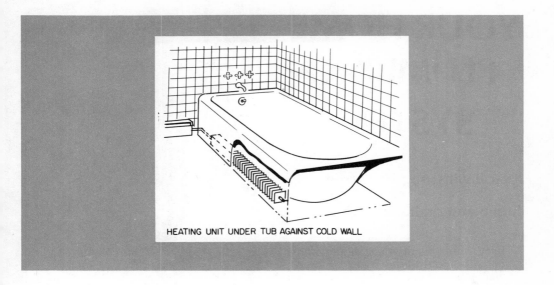

HEATING UNIT UNDER TUB AGAINST COLD WALL

borne dust and 99 percent of the pollen passing through the system. Most of these impurities in the air bypass a furnace air filter only to be recirculated.

Most electronic air cleaning units have screens to filter out the large particles of dirt. One has two treatment sections to trap tiny particles electronically. The first stage or ionizing section electrically charges the particles passing through it. The second stage or collecting section uses electrically charged metal plates that work like magnets to attract charged particles entering from the first stage. Material collected by an electronic air filter must be cleaned out every so often. Some units do this automatically. Others are manually cleaned.

An electronic air cleaner is usually installed in the return-air duct of a furnace. For houses without forced-air heating, individual room units are available.

OTHER ACCESSORIES

Outside air vent — A small outside air duct piped into the return-air side of a forced-air furnace makes a tremendous difference in the air freshness inside a house. A damper controls the flow of fresh air. A grille keeps out water and vermin. The slight positive air pressure

it produces counteracts infiltration of cold air from outside.

Instant hot water — Hot faucet water at the turn of a tap can be supplied by water heating coils built right into the boiler in a hydronically heated house. This eliminates the cost of a separate water heater and saves floorspace.

Automatic snow-melting — A system of tubing embedded in driveway and sidewalk concrete circulates heated water and antifreeze. The heat melts ice and snow accumulations. Heat is supplied by the boiler in a hydronically heated house. A swimming pool can be heated this way, too.

Heavy additional heating loads, such as snow-melting, must be figured when boiler size is calculated. Swimming pool heating, since it's off-season, need not be taken into account.

Heated bathtub — Whether you have dry or wet heat, as they term warm air and hot water heat, you can supply a little extra heat underneath the bathtub to warm it. It will seem more friendly when you shower or bathe. With dry heat a heating duct is run to the under-tub area and a vent cut in the subfloor to permit full air circulation. With hot water heat a small radiator is installed under the tub.

YOUR HOME PLUMBING SYSTEM

Basically, it's supplying water to the home and draining all the waste.

Every home handyman should know enough about plumbing to: (1) make simple plumbing repairs: (2) know what to do in an emergency; (3) be able to plan plumbing improvements; and (4) know when to call a plumber. If you can actually do your own plumbing around the house, you're one up on many handymen.

Whether you should try to do your own plumbing — the actual installation of pipes, fittings and fixtures — depends on your ability and your local code. Even if you know how, some codes won't let you work on plumbing. Only a licensed

You can do your own plumbing, even for a project as complex as a home remodeling.

plumber may be permitted to work. Such anti-do-it-yourself codes are prevalent in big cities where plumbers' union influence is strong. If you don't understand the principles of plumbing, you're better off hiring a pro anyway.

We're not recommending that everyone try to be his own plumber. It isn't everybody's bag. However, many handymen have done successful plumbing. You probably can, too. Be sure to read up on how to do a job correctly before you tackle it.

By doing your own plumbing, you stand to save up to 50 percent of the cost of having it done. And you can't put a price on the feeling of accomplishment when you turn on the water and watch it flow out of your pipe or your fixture. It works because you made it work. For jobs you don't do yourself, call in a plumbing contractor. He is skilled in

Local plumbing codes specify materials, sizes and how much of the job you can do.

planning, servicing and installing all your water needs.

Whatever you do, you're bound by your local plumbing codes. Nearly every city has them. If you live outside the city in a rural area, your county plumbing code applies to what you put in. You can get a copy of the code at your municipal building or county courthouse. Some areas have no plumbing codes at all. In this case, for your own protection, follow the National Plumbing Code. Then you'll be sure of ending up with a safe, workable system. Most public libraries have a book outlining the national code.

Codes likely will specify the following: the kind of pipe that's permitted for each use—drainage, water supply, vent, buried underground, etc.; venting requirements; fixture waste connections; trap requirements; and minimum pipe sizes for each use.

Your home plumbing system begins at the water utility's main, usually in front

PLUMBING SUPPLY SYSTEM. A. Source of water, public or private, and piping up to house. **B.** Stop and waste valve should be at the low point of the whole system. **C.** Cold water main is any line serving two or more fixtures. **D.** Hot water main is any line serving two or more fixtures. **E.** Branch is any line serving just one fixture. **F.** Shut-off valve is needed in every branch line and in mains where cutoff might be needed. **G.** Use air chambers at every branch line before the fixture to prevent water hammer. **H.** Fixture supply pipe is part of the branch line that fits it to the fixture. Study the drawing. **DRAINAGE SYSTEM. 1.** Fixture drain incorporates a trap and leads into the branch waste. **2.** Branch waste runs between the fixture and the main drain. **3.** Main drain, or soil stack, collects water from the toilet and branch wastes. **4.** Vent is the upper portion of the main drain. It reaches up through the roof. **5.** Revent is a bypass for air between a branch waste and the vent portion of the main drain. **6.** Cleanout opening should be located wherever access to the drainage system may be needed to rod out blockage. **7.** Building drain leads from the main drain to the point of final disposal. **8.** Final disposal is either a public sewage plant or a private disposal system.

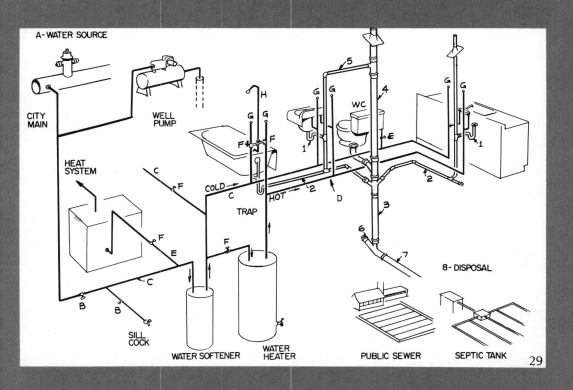

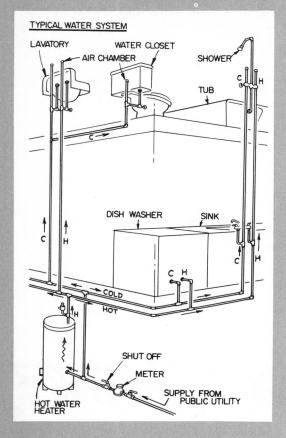

TYPICAL WATER SYSTEM

LAVATORY
AIR CHAMBER
WATER CLOSET
SHOWER
TUB
DISH WASHER
SINK
C H
COLD
HOT
SHUT OFF
METER
SUPPLY FROM
PUBLIC UTILITY
HOT WATER
HEATER

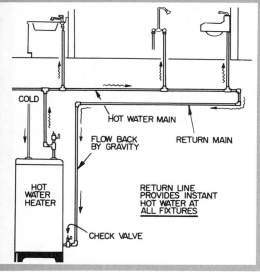

COLD
HOT WATER MAIN
FLOW BACK
BY GRAVITY
RETURN MAIN
HOT
WATER
HEATER
RETURN LINE
PROVIDES INSTANT
HOT WATER AT
ALL FIXTURES
CHECK VALVE

To avoid water hammer, water supply lines must have 12″ air chambers near fixtures.

of your house. If yours is a private well, it starts at the well. Plumbing is supposed to bring in all the potable water you need, treat it if necessary, heat some of it and deliver the hot and cold water around to where it's used. Water that isn't consumed must be drained safely out of the house by the plumbing. All of this must be done quietly and without danger to you or others in your community. It's a big job.

The average home has more than 300 feet of plumbing pipes in its walls, floors and buried in the ground around it. Basically, every home plumbing system comprises three separate parts: water supply system, fixtures and drainage system.

WATER SUPPLY

The water supply system consists of a service entrance line from the water main to the meter (with city water), a main shutoff near the meter, a distribution system, a water heater and perhaps a water softener and other treatment equipment. Room shutoffs and fixture stop-cocks also are included in the water supply system. Some fixtures use only one kind of water, hot or cold. For instance, most

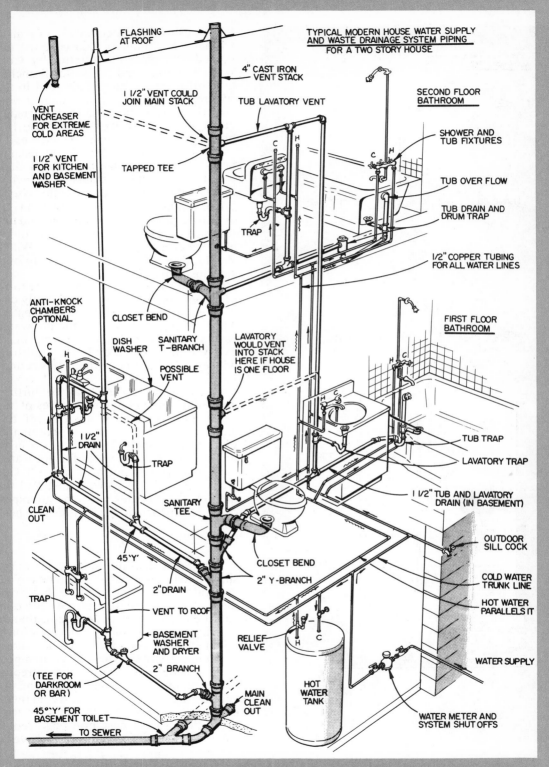

FLASHING
AT ROOF

TYPICAL MODERN HOUSE WATER SUPPLY
AND WASTE DRAINAGE SYSTEM PIPING
FOR A TWO STORY HOUSE

4" CAST IRON
VENT STACK

1 1/2" VENT COULD
JOIN MAIN STACK

TUB LAVATORY VENT

SECOND FLOOR
BATHROOM

VENT
INCREASER
FOR EXTREME
COLD AREAS

SHOWER AND
TUB FIXTURES

1 1/2" VENT
FOR KITCHEN
AND BASEMENT
WASHER

TAPPED TEE

TUB OVER FLOW

TUB DRAIN AND
DRUM TRAP

TRAP

1/2" COPPER TUBING
FOR ALL WATER LINES

ANTI-KNOCK
CHAMBERS
OPTIONAL

CLOSET BEND

DISH
WASHER

SANITARY
T-BRANCH

POSSIBLE
VENT

LAVATORY
WOULD VENT
INTO STACK
HERE IF HOUSE
IS ONE FLOOR

FIRST FLOOR
BATHROOM

1 1/2"
DRAIN

TRAP

TUB TRAP

LAVATORY TRAP

CLEAN
OUT

SANITARY
TEE

1 1/2" TUB AND LAVATORY
DRAIN (IN BASEMENT)

45 'Y'

OUTDOOR
SILL COCK

2"DRAIN

CLOSET BEND

2" Y-BRANCH

COLD WATER
TRUNK LINE

TRAP

VENT TO ROOF

HOT WATER
PARALLELS IT

BASEMENT
WASHER
AND DRYER

RELIEF
VALVE

WATER SUPPLY

(TEE FOR
DARKROOM
OR BAR)

2" BRANCH

MAIN
CLEAN
OUT

HOT
WATER
TANK

45°'Y' FOR
BASEMENT TOILET

TO SEWER

WATER METER AND
SYSTEM SHUT OFFS

31

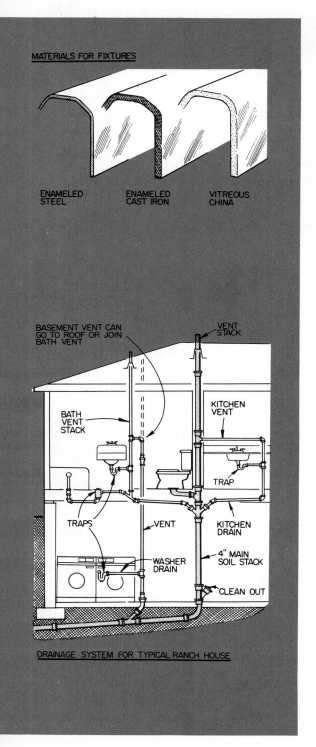

MATERIALS FOR FIXTURES

ENAMELED
STEEL

ENAMELED
CAST IRON

VITREOUS
CHINA

BASEMENT VENT CAN
GO TO ROOF OR JOIN
BATH VENT

VENT
STACK

BATH
VENT
STACK

KITCHEN
VENT

TRAP

TRAPS

VENT

KITCHEN
DRAIN

WASHER
DRAIN

4" MAIN
SOIL STACK

CLEAN OUT

DRAINAGE SYSTEM FOR TYPICAL RANCH HOUSE

toilets use cold water only; most dish-washers, hot only. In that case the un-tapped main passes that fixture by. Cold water branches and hot water branches take off from the mains and lead to the fixtures. Branches are of a smaller pipe size than mains because they normally carry water for only one fixture. If a branch serves more than one fixture, it's sized a little larger. Mains are usually held to the same size throughout their length. The parts of mains or branches that go up through walls are called *risers*.

Cold water mains may be split and run as double pipes to keep softened and un-softened cold water separate. Sometimes hard water is furnished to toilets. Water should always be furnished hard to out-side hose bibs.

The hot water main may be in the form of a continuous loop that constantly cir-culates hot water from the water heater. Then, no matter how long the hot water run, you'll have instant hot water at every tap. A large or long, rambling house should not be without this refine-ment.

A hot water system may be further divided to serve two temperatures of hot water. One is supplied at about 180 de-grees for dishwashers and automatic washing machines. The other is a mix-ture of hot and cold water for hot water taps. This comes out at a safe 120 de-grees. The mixing valve is installed on the hot water heater and separate pipes carried from there to the fixture branches.

FIXTURES

Fixtures such as tub, shower, lavatory, sink, toilet, dishwasher, etc., help you use water conveniently. More than any-thing else in plumbing, fixtures come in varying qualities. A good rule-of-thumb is to buy the best quality you can afford.

Three materials, chiefly, go into most plumbing fixtures: enameled stamped steel, enameled cast iron and vitreous china. Stamped steel tubs, sinks and la-vatories are cheap and tinny. Avoid them except where money is the only govern-ing factor. Either cast iron or vitreous china is very serviceable. Both have a solid feel, a "ring" as you tap on them. Some people rate china as best. It is the most expensive. All toilets are made of china. Other fixtures may be made from any of the three materials. Dishwashers and other appliance-type fixtures, of course, are made in stamped steel only. Nothing wrong with them because of that.

DRAINAGE SYSTEM

Used water that runs out of fixtures has to be carried away and disposed of. Gases created by decomposition within the system must be dispelled where you won't have to breathe them. This is the job of the drainage system. It must have pipes of ample capacity, properly pitched to carry wastes away by gravity. The system must be tightly sealed and properly vented. Adequate provision is needed for cleanout, should the drain pipes ever clog. The whole plumbing system is designed around the drainage system, since its pipes are the largest, most costly and most difficult to install.

The large pipe that runs vertically and collects wastes from one or more fixtures is called a *soil stack* or just *stack*. If a toilet empties into the stack, it's called a main soil stack. Every home has at least one main stack. There may be more if required to serve more than one toilet.

If no toilet drains into a stack, it can be of smaller pipe than a main stack. This is called a secondary stack. Each fixture is joined to its stack by a branch drain. The branch drain must slope downward toward the stack. A stack extends from its above-roof vent to below the house. There it connects with a horizontal sloped run that's called a *house drain*. Ordinarily there is only one house drain, although there may be several. The house drain becomes a *house sewer* as it leaves the house and enters the ground outside the foundation. The house sewer connects with the city sewer or with a private sewage disposal system.

Additional pipes serve gas only, no drainage. These are called *vent pipes*. The main vent pipe, of course, is the por-

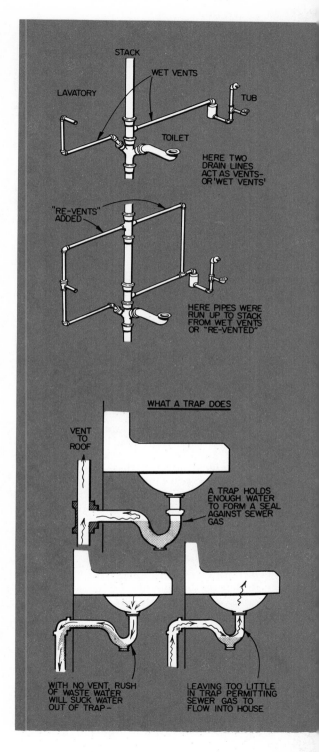

STACK
WET VENTS
LAVATORY
TUB
TOILET

HERE TWO DRAIN LINES ACT AS VENTS— OR 'WET VENTS'

"RE-VENTS" ADDED

HERE PIPES WERE RUN UP TO STACK FROM WET VENTS OR "RE-VENTED"

WHAT A TRAP DOES

VENT TO ROOF

A TRAP HOLDS ENOUGH WATER TO FORM A SEAL AGAINST SEWER GAS

WITH NO VENT, RUSH OF WASTE WATER WILL SUCK WATER OUT OF TRAP—

LEAVING TOO LITTLE IN TRAP PERMITTING SEWER GAS TO FLOW INTO HOUSE

Hot water tank mixer provides 2-temperature water. Center pipe carries 140° mixed hot and cold water; pipe at left 180°.

WATER HEATER ON HEATING SYSTEM BOILER

STORAGE TANK

INDIRECT HOT WATER HEATER

WATER IN

WATER LEVEL

WATER COIL

WATER OUT

STEAM OR VAPOR HEATING BOILER

tion of the soil stack that extends above the highest fixture and up through the roof. Fixtures may be vented into the soil stack through their waste pipes if they are close enough to the stack. This is called *wet venting*. Under the National Plumbing Code, a tub or shower within 3½ feet from trap outlet to soil stack qualifies. So does a lavatory within 2½ feet of the stack. If the fixtures are farther than this from the stack, they need separate pipes running from the fixture, either up through the roof or horizontally and connecting into the soil stack above the highest fixture drain connection. These "vent-only" runs are called *revents*. Horizontal revent pipes are pitched slightly upward from the fixture to make condensation drain back toward the fixture.

Drain and vent pipes are sized according to how much fluid each must carry. More about that in a later chapter. Toilets are hooked up to the soil stack by a large pipe called a *closet bend*.

Every toilet has its own built-in water trap. All other fixtures must be provided with separate traps to keep gases and vermin in drainage pipes from escaping into the house. A trap is a U-shaped pipe that is always filled with water. The water seals off drain piping beyond it.

WATER HEATER

An important part of any plumbing system is the water heater. The most common types are fueled by gas, electricity or oil. They can't heat water as fast as it is used, so some already-hot water must be stored inside the heater's tank.

Another type of heater that is available to those with hot water house heating systems is the instantaneous type. Water to be heated is run through coils that are immersed inside the heating plant's boiler. There the water is warmed to

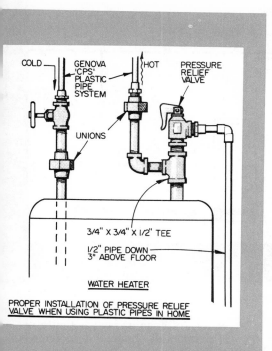

COLD

GENOVA 'CPS' PLASTIC PIPE SYSTEM

HOT

PRESSURE RELIEF VALVE

UNIONS

3/4" X 3/4" X 1/2" TEE

1/2" PIPE DOWN — 3" ABOVE FLOOR

WATER HEATER

PROPER INSTALLATION OF PRESSURE RELIEF VALVE WHEN USING PLASTIC PIPES IN HOME

about the same temperature as the boiler water. Sometimes a storage tank is used with this kind of heater, too. One drawback, the boiler must be kept up to full temperature at all times if hot water is wanted. This may not be too practical in any but the coldest weather.

You can judge the quality of a fueled hot water heater pretty well by the guarantee offered and the reputation of its manufacturer. Since the guarantee represents the manufacturer's confidence (or lack of it) in his product, study it carefully. There are many "10-year guarantees" that give you only five years of full-value guarantee plus five years of pro-rated guarantee. The good guys offer you a full-value guarantee for the whole 10 years. Then if the tank should leak before the 10 year guarantee period is up, you'd get another water heater free.

As for reputation, a number of heater manufacturers have good names in the field.

What size heater? Your family's need for hot water varies from day to day and year to year. However, a 20-gallon tank is the minimum recommended for any heater using gas or oil as a fuel. A family of four should have at least a 30-gallon heater. Larger families and those with automatic laundry equipment or automatic dishwashers need at least a 50-gallon heater in gas or oil.

How much hot water your family uses depends upon how many water-wasters you have. Normally a tub bath takes 10 to 15 gallons, a shower 9 to 12 gallons. An average load of clothes in an automatic washer uses 13 to 20 gallons of hot water, plus what's needed for rinsing. Hand-dishwashing uses 2 to 4 gallons. Automatic-dishwashing takes 7 to 12 gallons. These amounts vary depending on the equipment and the temperature of the water.

The cost to heat water depends on the cost of the fuel used. If you know the fuel prices in your area, you can figure it out. Here is a reasonable estimate of the amounts of each fuel used per year: Natural gas — 32,450 cubic feet; manufactured gas — 61,800 cubic feet; LP-gas (bottled — 1475 pounds; oil — 270 gallons; electricity — 6658 kilowatt hours; and coal (some still use it) — 1.8 tons. For instance, if fuel oil in your locale costs 14 cents a gallon, your total water heating bill for the year would be $.14 times 270 gallons, or $37.

Never forget that a water heater is a potential steam boiler. If the temperature control fails to shut the burner off, the water will heat until it becomes steam. When enough steam pressure builds up, the heater will blow. People have been injured and killed this way, houses damaged. To be safe, every water heater should be fitted with a temperature-pressure relief valve with its heat-sensing probe reaching down into the tank.

If your temperature pressure relief valve has a pump-like handle (most do), test it to see that it's working by "popping" the valve and squirting off a little water.

Owners of newer water heaters are further protected by devices known as

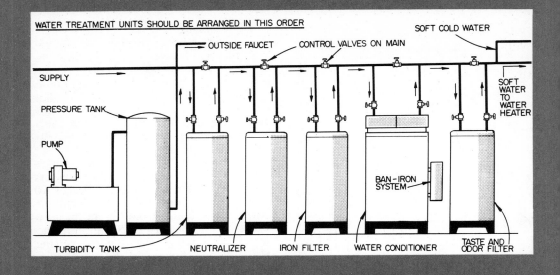

SOFT COLD WATER

OUTSIDE FAUCET — CONTROL VALVES ON MAIN

SUPPLY

PRESSURE TANK

PUMP

SOFT WATER TO WATER HEATER

BAN-IRON SYSTEM

TURBIDITY TANK — NEUTRALIZER — IRON FILTER — WATER CONDITIONER — TASTE AND ODOR FILTER

energy cutoffs. If the temperature should get too high, one of these safety controls would shut off the gas or electricity.

WATER TREATMENT

Water pumped out of deep wells is often hard. This means that it contains dissolved calcium and magnesium sulfates or bicarbonates. It may also contain iron and other minerals.

The surest way to tell about your own water hardness is to have it tested for mineral content. The major water softener manufacturing firms, such as the Lindsay Company, will make a free test of your water. They can then tell you what size of softener you need. Large mail order houses, like Wards and Sears, do this too.

Most softeners remove a small amount of iron and sulfur while softening the water. But when the content of these impurities gets too high, additional water-conditioning equipment may be needed ahead of the softener.

Bad-tasting water can sometimes be improved by running it through a charcoal filter. Small charcoal filters are available for insertion in the line at the faucet where drinking water is usually drawn.

A water softener doesn't actually remove the minerals. Instead it exchanges sodium ions for hard calcium and magnesium ions. When all the sodium ions are used up, they're replaced by running a salt solution through the softener media. Some equipment does this automatically by running water from a brine tank through the softener. In others you dump the salt in and flush it through with water.

Softeners are available for outright purchase or from some firms on a rental basis. Although it's less trouble to rent, in the long run it's more economical to buy a softener.

Plumbing improvements can be financed through your bank, savings and loan association or other private lender. They also qualify for lower-interest FHA home improvement loans. Time payment plans by those who sell plumbing supplies are available at higher interest. It pays to shop around for the best deal on getting your money.

If you are thinking of buying a house, check out its plumbing system to see that you're not buying trouble.

1. Look at the fixtures closely. Is their style in keeping with the rest of the house? Are they made of enameled stamped metal, enameled cast iron or vitreous china? Are they chipped or

scratched?

2. Fill up the lavatory and bathtub with water. Is it clear or murky? Do the stoppers hold water? Do the fixtures drain quickly and quietly?

3. Flush the toilet with a small ball of crumpled-up toilet paper or cigarette butt in it. Is the flush complete? Does the toilet tank refill and shut off fully? How noisy is it?

4. Run the water full blast in the kitchen, then turn it on in the bathroom. Does a good flow come out or just a trickle?

5. Open and close each faucet several times to see whether it leaks.

6. Are there rust stains in the bathtub or lavatory? Lift off the toilet lid and look there for evidence of minerals in the water.

7. Get a look at the roof above the toilet. See that there is a vent stack and that it's away from windows.

8. Check the size of the water heater to see whether it's ample to supply hot water for your whole family. It should bear the label of the National Board of Fire Underwriters and be fitted with a relief valve.

OUTSIDE YOUR HOUSE

Your municipal water supply and sewage disposal may be considered as extensions of your home plumbing system. Water that has been drawn from a surface or underground source sometimes must be treated to make it fit to drink. Raw water may be precipitated, settled, filtered, softened and chlorinated before it's fit. Water's alkalinity-acidity must be adjusted. *Finished water*, as it is called, is stored in reservoirs until needed, then pumped through underground water mains to your house.

Constant checks are made on city water for health reasons. All this is the responsibility of the governmental agency or utility that furnishes your water.

After water leaves your house as sewage, it travels by gravity, usually, through sewer pipes to a sewage treatment plant. There solids are settled and filtered out. Oxygen for bacterial action is provided by aeration. The resulting effluent is

A water softener needs to be regenerated regularly. It's simply a matter of adding salt, thereby providing new sodium ions.

Potassium permanganate is sometimes added to water in iron filter to replace used-up oxygen. Flush out before using filter.

chlorinated to kill any remaining bacteria and render it harmless to a river or stream where it is discharged. You've heard the story about the sewage treatment plant operator who drinks a glass of fully treated sewage effluent at the close of a plant tour to show how pure it is. Now *that* takes guts, but it can be done.

CROSS-CONNECTIONS

One thing you should watch out for in any plumbing is a cross-connection. Cross-connections can be murder. You won't believe this, but only one house in thousands is completely free of them. A cross-connection can exist for years without harm. Then, when conditions are right — wrong, actually — it lets pollutants get into your potable water. There is no warning. The first you know, someone is sick. Death can even result.

To keep your family out of all possible danger, make a cross-connection inspection right now.

A cross-connection is a link or channel between a pipe carrying polluted water and a pipe carrying drinking water.

When pressure from the polluted source exceeds pressure on the drinking water, the contaminants enter the potable water system through the link. This action is termed backsiphonage or backflow. It's simply a reversal of water pressure. Backflow can come about in a number of ways. It can happen when water in the house system is turned off, when a car hits a fire hydrant, when emergency use of city water forces a drop in pressure, or

in any number of other unusual circumstances. Here's another: In a system where the pipes are poorly sized, a full flow in one pipe, such as might be caused by opening a faucet, can create a vacuum in another pipe. None of these situations is impossible.

Through siphon action, contaminated water can not only flow downhill, it can flow over a "hill" created in house piping. Water can actually be drawn up into your house water pipes. When normal flow is restored, the contaminated water may flow out of any outlet.

Finding cross-connections takes a little doing. Correcting them takes real de-

Tapping test tells if a fixture is low-cost stamped metal or heavier cast iron or vitreous china. You can recognize enameled metal by its tinny sound.

To test a home's water quality, run out a tubful of hot water to see how clear it is. Coins on the bottom should be bright, easily seen through water.

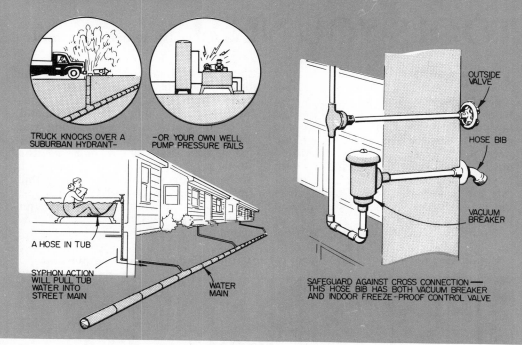

TRUCK KNOCKS OVER A
SUBURBAN HYDRANT—

—OR YOUR OWN WELL
PUMP PRESSURE FAILS

OUTSIDE
VALVE

HOSE BIB

VACUUM
BREAKER

A HOSE IN TUB

SYPHON ACTION
WILL PULL TUB
WATER INTO
STREET MAIN

WATER
MAIN

SAFEGUARD AGAINST CROSS CONNECTION—
THIS HOSE BIB HAS BOTH VACUUM BREAKER
AND INDOOR FREEZE-PROOF CONTROL VALVE

termination. The link may be a lot more subtle and harder to notice than a direct pipe connection. The factors that can produce a reversal of flow may seem remote or even impossible. Yet the siphon principle is always ready to take over any time there's a vacuum in the system.

When making your inspection, look for two types of cross-connections: (1) a solid pipe with a valved connection and (2) a submerged inlet.

The solid connection is installed where necessary to supply an auxiliary piping system from the potable water source. The pipe supplying water to a hot water heating system is a common cross-connection. Most codes permit it. It's a cross-connection that many families live with every day. Whether your family should or not is up to you. A backflow-preventer in the line would head off trouble.

Submerged inlets can be found in old-fashioned lavatories, sinks and bathtubs built before the danger of cross-connec-

tions was recognized. Some newer fixtures have them too. It's possible that your toilet tank inlet is submerged. This is another cross-connection most of us live with.

Follow your water pipes from where they leave the water meter or pump to the end of every branch line. Imagine drinking the water in everything they're connected to.

A laundry tub, or pail, with a hose in it is a cross-connection. Disconnect hoses when not in use, or else provide a vacuum-breaker at every hose connection.

A rain water cistern may have a valved pipe cross-connected to the potable water. Remove the pipe.

See that an in-the-ground lawn sprinkling system is installed with a vacuum-breaker that's at least 12 inches above the highest sprinkler outlet.

The U.S. Public Health Service says that taking *any* chance with family or community health is too much.

TOOLS YOU NEED   These are basic tools

Plumbing and heating construction and repairs go easier if you have the right tools. More than likely you already own many of the ones you'll need. They are basic to any home maintenance. You can get the others as you need them. If the need is only occasional or the tool is costly, rent it instead of buying. As with anything else you buy, there's no substitute for quality in a tool. Quality costs more, of course, but the satisfaction and long life are well worth it.

No matter what kind of work you do, there are certain basic tools you should have. You'll need a claw hammer, ball peen hammer and 6-foot or longer measuring tape or folding rule. A folding rule is handiest for heating-plumbing runs because it is rigid. Also included in the list of basic tools are 6- or 8-inch slip-joint pliers, locking plier wrench, straight-blade screwdriver, No. 2 Phillips screwdriver, 6- and 12-inch adjustable open-end wrenches, metal or aircraft snips, hacksaw, keyhold saw, plumber's chisel and cold chisel.

All-purpose power saw handles bigger cutting jobs. Blade gets into corner areas.

For drilling, you'll need a set of drill bits and a set of wood-boring bits of one kind or another. The wood-boring bits, in steps from ¼ to 1 inch, are for use with an electric drill. You'll need that too. A ¼-inch drill is big enough for small jobs. If you do lots of piping, such as for a house addition, it will pay to rent a ½-inch electric drill with anglehead. This is used with reworked auger bits for production hole-drilling. The bits are remodeled by sawing off their shanks to leave just the round portion. Chucked up, they feed themselves into the wood, saving you much effort. It's a trick professionals use.

PLUMBING TOOLS

No matter what work you do in plumbing, you are confronted with the need for some basic plumbing tools:

Pipe wrenches—You need a pair, one to turn, another to hold. A useful combination is a 14-inch and an 18-inch. Choose either the standard or Stillson pattern.

Other pipe wrenches are made for special applications. You may or may not need them. They're in the handy-to-have category. One type is made for better gripping on hex nuts, such as those found on threaded steel unions. It comes with either straight or offset jaw. Likewise, standard pipe wrenches are available with offset jaws for working close to a wall or a parallel pipe. Still another type has its jaw opening parallel to the handle for working tight up to a wall. These special pipe wrenches cost somewhat more than standard ones.

A chain pipe wrench gives you a tight grip on all shapes without crushing. It works well with large pipe sizes up to 4 inches and more. It also works in tight places. A lower cost strap wrench grips like a chain wrench but uses a strong cloth strap in place of the chain. A strap wrench is suited to turning polished pipes without marring the finish.

Monkey wrench—This utility wrench is especially useful beyond the sizes a large adjustable open-end wrench will fit. It's good for tightening slip-nuts on drains and traps.

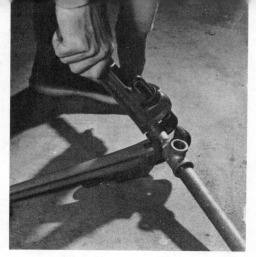

Photograph, above, shows a standard pipe wrench. This tool is basic to plumbing.

With many holes to bore, rent ½″ electric angle-head drill. Use cut-off auger bits.

Left to right: flat bit, better auger bit, and best screw-in auger with a cut-off shank.

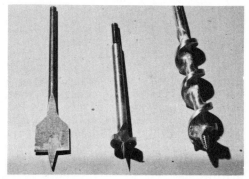

If wood rule isn't handy, use steel rule.

Use hex wrench for work on hexagonal nuts.

Use offset wrench for pipes next to walls.

Large pipes take heavy-duty chain wrench.

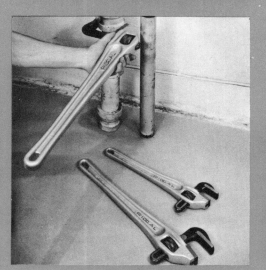

Basin wrench — Made for turning faucet hold-down nuts and supply pipe collars behind a fixture bowl, this highly specialized tool is indispensable in installing an inplace lavatory or sink. You'll have real trouble tightening the fittings any other way. A basin wrench has a swivel jaw that tightens or loosens, depending on which way it's swiveled.

Hole-saw — Used in combination with an electric drill, a hole-saw takes over where clean holes larger than 1 inch are needed. It's much faster and neater than drilling a circle of smaller holes and knocking out the remaining plug.

Saber saw — This is a great tool for cutting heating register outlets from floor or wall surfaces as well as for obtaining clearance for other heating and plumbing parts that need more than a round hole. There's a larger *all-purpose saw* that is a he-man saber saw, you might say. It's most useful for remodeling framing to accommodate plumbing or heating runs.

STEEL PIPE

Here are special tools you need for

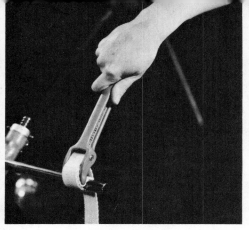

Strap wrench turns pipes without damage.

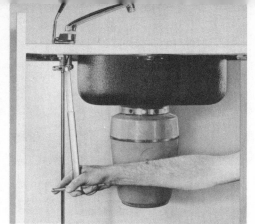

Ridge Tool Co.

This basin wrench can grip and turn nuts in almost impossible areas behind sinks.

handling steel pipe.

Pipe cutter — This is for cutting steel pipe prior to threading it. It's easier to use and does a better job than sawing.

Pipe reamer — After the pipe has been cut, a burr is left around the inside. This is removed with a pipe reamer.

Pipe vise — If you thread your own steel pipe, you'll need a pipe vise to hold the pipe while you thread. One especially handy vise clamps to a bench or vertical support. All have yokes that swing out of the way for getting a pipe in and out of the vise quickly and easily. If you buy a pipe vise, get a size that will handle the largest pipe you'll want to thread.

Pipe threaders — Pipe dies make tapered threads. Even though you have a stock and dies for bolts, you'll need an-

other set for pipe threading. The handiest ones come with ratchet handles. Order separate die head for each size of pipe. Make sure the stock will accommodate the largest size you'll be using. Ratcheting stock-die sets often come with three dies for threading ½-, ¾- and 1-inch pipe.

COPPER PIPE

When working with copper pipe, your tool needs are simpler than with steel pipe.

Tubing cutter — Like a pipe cutter but smaller, a tubing cutter leaves a square end on copper pipe. Burrs on the inside should be removed, it will restrict flow through the pipe. Most tubing cutters

Hole-saw bores up to 2½″ in floors, walls.
Stanley Tools

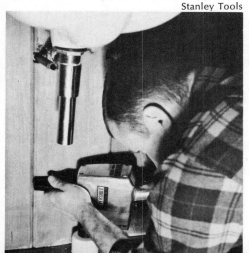

Pipe vise opens and closes for fast work.

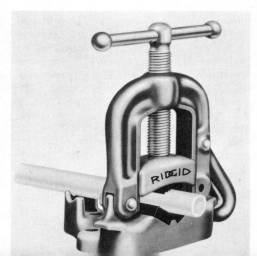

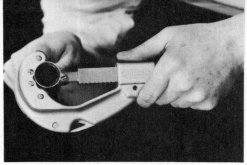

Ridge Tool Co.

Tubing cutter has ratchet for adjustment.

Use a high-heat propane torch for sweat-soldering DWV and other copper fittings.

Some sewer tape heads are interchangeable.

also have compartments for extra cutter wheels. You'll pay several dollars or more for a good one.

Propane torch — For sweat-soldered copper pipe joints you'll need a propane torch. You could use a gasoline blowtorch, but the propane one is much more convenient to use. Safer and cheaper too. For soldering on DWV pipes, get a high-output torch head. Bernzomatic makes one. Another nice touch is a two-stage flame: pilot for standby and full flame for soldering. Keep a striker with your torch to light it without matches. Much handier.

Flaring tool — If instead of soldered joints, you use flared joints, get a flaring tool. There are two types: One clamps the pipe in a hole while a metal die is screwed against the end to flare it. The other type consists of just the die, which is pounded into the end of the pipe for flaring. The screw-in kind will handle many sizes.

CAST IRON SOIL PIPE

When you work with cast iron pipe you'll want to have a few specialized tools:

Calking tool — This chisel-like affair is useful for forcing oakum into the joints and for dressing off the hardened lead to ensure watertight calked joints. The tool

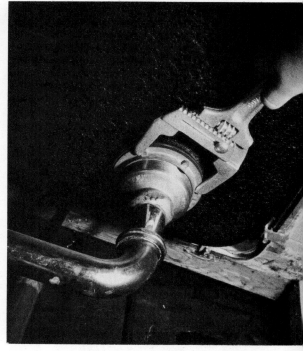

Self-storing auger can be worked through the roof vent to unblock stuffed pipes.

Use a special wrench to tighten plated sink and lavatory fittings without damage.

is merely an offset flat-ended chisel.

Ladle — For melting and pouring lead into calked soil pipe joints, a ladle lets you hit where you aim. You can get by with an empty tin can held with pliers, but then you stand a good chance of getting burned.

Asbestos joint-runner — This tool is a "must" item for leading horizontal soil-pipe joints. The asbestos ring fits around the outside of the joint with just enough room to pour molten lead in at the top. The ring seals the molten lead from leaking out.

PLUMBING MAINTENANCE TOOLS

Plumber's friend — What home is without this ubiquitous tool? You may have to experiment a bit to get one that works on every toilet bowl. Smaller sizes fit lavatory and sink drains better and cost less that the full-sized ones.

Sewer tape — Made in 50-foot lengths

and heavy-duty 100-foot lengths, the sewer tape is rammed into a drain pipe to unclog it. A good practice is to keep a 50-foot tape on hand and rent a heavier one if you need it.

Closet auger — This is a flexible-shaft auger just long enough to reach through toilet bowl passages and free them of obstructions. If you have small kids around who like to flush toys down the toilet, you definitely need a closet auger.

Self-storing auger — the self-storing sewer auger is handier and more effective than a sewer tape for clearing stubborn drains. Use it through the roof vent stack or any other cleanout opening. Such a sewer auger carries its 15- to 25-foot flexible cable in the crank handle. The cable is fed out as needed. Some have interchangeable end fittings for cutting, boring, breaking, sawing and chipping. This tool is the next thing to power sewer cleaning equipment. If the big stuff is needed, rent it.

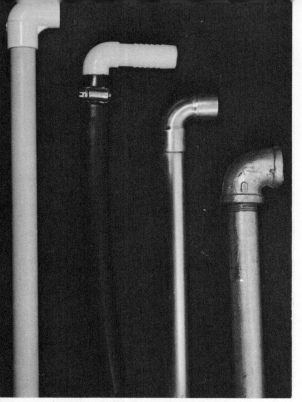

Plastic, flexible polyethylene, copper and steel are the types of pipe and fittings available for installing water supply lines.

WATER SUPPLY PIPES AND FITTINGS

Plastic pipes cost little and are also easy to use

You have a wide choice of materials to use in a home water supply system. Each one comes with its own way of joining the different pieces together to make runs of pipe. To some extent the materials you can use are limited by local code. Check yours before picking a pipe type.

Water supply pipes are made in plastic, copper and steel. Lead too, but forget it. Too hard to work with. Plastic pipe is the easiest to handle. Its smooth, almost frictionless walls pass water better than any other material. In use it has a few limitations. More about them later. Copper pipe is easy to handle and almost all codes okay it. Neither copper nor plastic pipe will corrode even in very hard water. Steel pipe, because of its serviceability and low cost, is still the standby of many plumbers. It costs more than the others if you figure the time it takes to cut and fit the threaded joints. While you can work with steel pipe successfully, and it's a challenge, there are easier materials to use.

Hot and cold plastic water pipe is the newest thing in plumbing. The standby polyvinyl chloride (PVC) pipe has been chemically toughened to the point where it can be used for both cold and hot water supply lines. The new pipe is called *chlorinated polyvinyl chloride,* CPVC for short. Part of the family of rigid PVC materials, its sister pipes have seen years of service in cold water and drainage-waste-vent installations. CPVC is an improved version of what used to be termed *PVDC,* polyvinyl dichloride. The older material, with threaded joints, was used for hot and cold water supply in houses as early as 1960. The recent improvements should make it the plumbing of the future. More than 100,000 installations are now in service. CPVC pipe is rated to take 100 pounds per square inch (psi.) pressure at 180 degrees.

CAN YOU USE IT?

Plumbing and codes are synonymous, practically. And plastic pipe isn't mentioned in most codes. The reasons vary from "let's wait and see what the others do" to "we've never done it that way before." In any case you are left to interpret your code's provisions relating to plastic piping. Since, good or bad, you are bound by local codes, check yours before going ahead.

I know a man and his teen-age son who installed CPVC piping throughout their

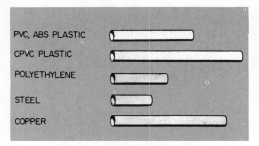

PVC, ABS PLASTIC	
CPVC PLASTIC	
POLYETHYLENE	
STEEL	
COPPER	

Water supply valve engineering

new home a few years ago. Upon completion, the system was pressurized to 100 psi. and the pipes left under pressure overnight. In the morning the pressure gauge still showed the original 100 psi. No leaks. After three years of use, there still are no leaks and no other problems either, they report.

A remodeling contractor I know will use no other piping material because of the ease of working with plastic. In three years of using CPVC, he has never had a problem or complaint.

I interviewed other handymen who have used the new plastic pipe in home plumbing additions and modernization work. All are completely sold on it; none reported any troubles with it.

Plastic pipe weighs only 1/20th of what galvanized pipe of the same size

does. It's lighter than the lightest copper pipe, weather-resistant, too. You can use it outdoors. You need no flame, no flux, no hot solder. Another advantage, plastic pipe doesn't sweat as readily as metal pipes. Neither does it feel blistering hot when carrying hot water. The self-insulating benefits of plastic will help you save on hot water costs by cutting heat loss in pipes.

No taste or odor is passed into the water. Plastic pipe is available with all the fittings you'll need for a complete installation. These include transition fittings, elbows, tees, wyes, couplings, reducing fittings, traps, saddle connections and caps.

DRAWBACKS

Anything this good must have some drawbacks. CPVC pipe is no exception. You'll have to be careful with your plastic pipe installation. Proper installation is no different for plastic pipe than it is for any other. It's just more important to the success of the system. The secret is to design the system to avoid temperatures of more than 180 degrees and pressures of more than 100 psi.

See that your water heater has a temperature-pressure relief valve on it. Its temp-relief setting may be as high as 210 degrees, but plastic pipe will take tem-

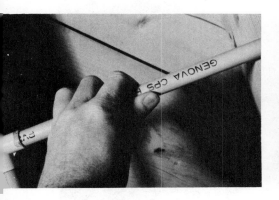

The type of plastic pipe in the above photo can carry cold water because it's self-insulating. It doesn't sweat like metal pipe.

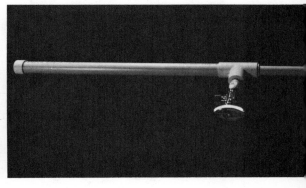

Keep pressures in plastic pipe down. This ¾" x 18" air chamber provides air cushion for quick-acting auto washer shutoff valve.

A 10 foot run of plastic hot water pipe can expand ½″ when it gets hot. Plastic pipe hangers are specifically designed to allow back and forth movement of pipe.

As shown in photo, flexible polyethylene pipe is held to fittings with screw-type clamps. Poly pipe is good for use in both wells and underground sprinkler systems.

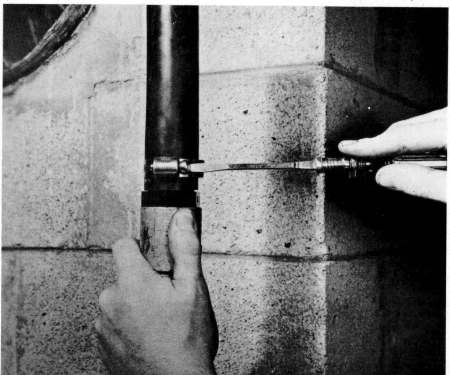

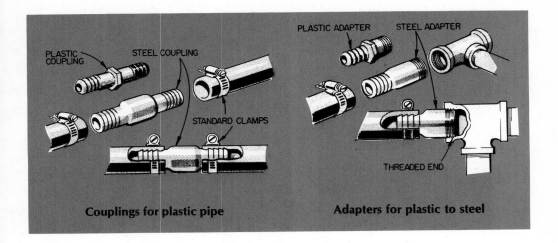

peratures over 180 degrees for only short periods.

One plastic pipemaker recommends water heater setting of 140 degrees but approves of settings as high as 180. This is hotter than you'd want water for safe use outside a dishwasher or washing machine.

Overpressures on pipes in a plastic system must be guarded against. If you turn on the water full blast, then turn it off quickly, whammo, water hammer. The fast-moving water slams against the closed valve and kicks up its heels in all directions. You've created an instantaneous pressure of perhaps 500 psi. Copper and steel pipe can take some of that. Plastic can't. Fast shutoff occurs regularly in washing machines and dishwashers with solenoid valves. Also with some of the new fast-acting faucet valves.

The solution is to see that your plastic plumbing system is fitted with 12-inch-long air chambers as close as practical to every water use. Such air chambers are standard in most plumbing. They're formed by capping lengths of pipe and extending them upward from tees at the stub-outs leading to each fixture. Often the air chambers for a solenoid-valve-controlled outlet are made 18 inches long and of one size larger pipe than the supply pipe.

Protect plastic piping with metal strips where it crosses studs so that nails don't get driven into it.

One last thing to keep in mind as you install plastic pipe is to allow for expansion and contraction of long, straight runs of pipe. A 10-foot length will expand ½ inch when heated from room temperature to 180 degrees. The pipe should be supported every 3 feet or less in hangers that permit linear movement and don't compress, cut or abrade the pipe. Use hangers designed for plastic pipe installation.

One other drawback, plastic piping cannot be used for an electrical ground. Find some other method.

Installed with good plumbing practice, CPVC plastic pipe is great. It's the best thing to hit plumbing since the flush toilet.

OTHER PLASTIC PIPES

CPVC isn't the only kind of plastic pipe you can use for a plumbing supply system. Rigid PVC and ABS (acrylo-mitrile-butadiene-styrene) can be used for cold water pipes, but not hot. These pipes come in 10-foot lengths with matching fittings. Since they cost only one-fifth as much as CPVC pipe, you might consider using one of them for cold water and CPVC for hot water. However, there's always the chance of a mix-up in fittings or pipe when running your hot water installation. The saving may not be worth that risk.

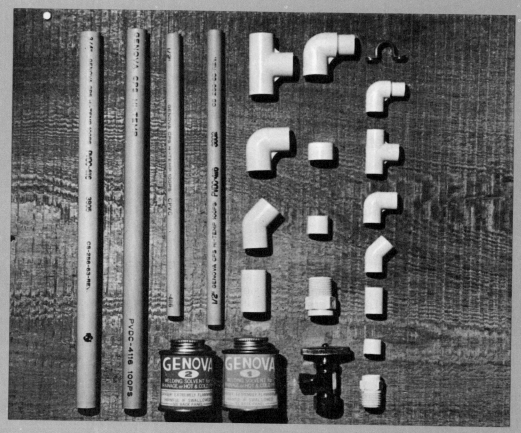

Pipe and fittings for a plastic water supply system let you build it the easy way. Included are: color-coded hot and cold pipes in ¾" and ½", tees, 90° street elbows, 45° elbows, couplings, ¾"-½" reducer, threaded adapters, joining cements for plastic pipe.

Polyethylene pipe is the flexible black pipe you may already be familiar with. Like PVC and ABS pipe, poly is to be used for cold water runs only. It's flexible, letting you negotiate gentle curves without the need for fittings. Poly pipe comes in 100-foot coils or longer and is cut to length with a sharp knife or hacksaw. There are various grades, from downright economical to more costly. Which of these to use depends on the strength needed. For water supply use 100 psi. pipe. For sprinkler systems you can use the cheaper 80 psi. pipe. Poly pipe is slipped over long, ridged plastic fittings and held in place with screw clamps. The joints can be taken apart and put together many times and remain leak-free. If a poly pipe resists pull-off after its clamp has been loosened, pour hot water over the end of the pipe to soften it.

The primary use of poly pipe is where lightweight, easy installation and flexibility are needed for a cold water run. Think of poly for a service entrance, well pipes down a well or a buried lawn sprinkling system.

JOINING PLASTIC PIPE

Plastic pipe can be installed with a few ordinary tools. You need a fine-toothed saw to cut it (9 to 14 teeth per inch) or a 24-tooth-blade hacksaw. You can cut plastic pipe with a power saw, too, if you like. If you use a vise to hold the pipe, wrap it with cloth to prevent damage to the pipe. You'll need a small knife or sandpaper to remove burrs from the insides of pipes after cutting.

Lastly, you'll need one or two non-synthetic bristle brushes with coarse bristles for applying joint solvents.

Plastic pipe will amaze you in its ease of installation. First cut it to length and remove the burr. Clean the pipe end and inside of the fitting. Next (with Genova pipe) brush the fluxing solvent from the first can on the outside of the cleaned pipe end for a distance equal to the depth of the fitting. Brush the inside of the cleaned fitting too. Don't brush the solvents out. For best results the brush used should be about half as wide as the pipe.

Immediately follow by liberally brushing the thicker solvent from the second can on the pipe end and a light but even coat in the fitting. Best results are had by brushing in the direction of the pipe and making sure that all mating surfaces are coated with cement.

Ideally, doping of both the pipe and fitting should be done in less than a minute. Without waiting, push the pipe and fitting together with a slight twist until the pipe bottoms in the fitting. Quickly adjust the fitting direction and that's it. Some instructions call for the fitting to be held on for 15 seconds. It's not necessary with CPVC pipe. Don't disturb the joint for about three minutes after assembly. The fitting will become immovable within seconds after insertion on the pipe.

Get it properly aligned without delay. This is the only hang-up of working with plastic pipe. If you blow it, the fitting must be sawed off and a new one used.

The CPVC tubing system is designed to have an interference fit. Check the fit before doping each joint by slipping the pipe into the fitting without a solvent. Proper fit is when the tubing makes contact with the fitting walls between one-

For a square cut in plastic pipe saw it in miter box with a 24-tooth blade in hacksaw. Remove burr on inside of pipe with penknife.

Coat outside of cleaned pipe and inside of plastic fitting with solvent. This Genova joining system uses 2 solvents; #1 and #2.

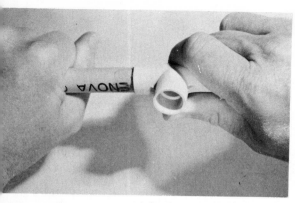

After coating, assemble joint with slight twist, getting pipe all the way into fitting. You must align it before joint sets.

third and two-thirds of the way in. Reject CPVC pipe and fittings without an interference fit.

Solvent cement that has jelled, discolored or noticeably thickened should not be used. Don't breathe the vapors of the solvents used to join plastic pipe. Work in a ventilated room and keep the solvent away from fires.

PRESSURE-TEST

The joints you produce in plastic pipe are stronger than the pipe itself. They're leakproof and will stand many pounds of pressure within a few minutes, but it's best to wait at least 16 hours before pressure-testing the system for leaks. A normal water pressure should be applied to the system. It should hold for about four hours in the closed system without falling off.

If a leak shows up during pressure-testing, that part of the system should be drained and the leaky joint and fitting cut out. Dry the pipe thoroughly and re-install a new fitting, using couplings and short lengths of pipe. These joints may be assembled without rotating if you are careful to coat all the mating surfaces uniformly and completely. Allow another 16 hours before pressure-testing the repair.

Don't try solvent-welding at temperatures below 40 degrees.

An epoxy-welding technique can be used to join CPVC pipe directly to copper fittings. Genova makes a "heatless welding kit" for this purpose.

Sizes of the new plastic pipes for hot and cold water application follow those of copper pipe in order to take advantage of the labor-saving compression fittings for lavatories and toilets that are now in stock at hardware and plumbing supply outlets. Be sure to replace the brass ferrules with special plastic ferrules designed for use with plastic pipe. CPVC pipe is easily available in ½- and ¾-inch sizes.

Plastic pipe can be joined to present copper water pipe with one simple compression fitting. To connect it to an existing steel pipe system, a threaded steel pipe adapter is all you need. Never thread plastic pipe. Use threaded adapters when threads are necessary. Don't overtighten threaded joints with plastic fittings. Get them hand-tight, then give not more than 1½ turns additional tightening with a strap wrench. Using a pipe wrench or chain wrench on plastic threaded adapters is a no-no. The teeth chew heck out of the fitting. Use a recommended pipe joint compound on all threads. Teflon tape is a good one.

Don't try to paint plastic pipe. To give built-in color-coding, Genova CPVC pipe is made in two colors—green for cold water lines and red for hot water lines. Other than color, there's no difference in the two. You can switch the colors if you need to fight the establishment.

COPPER PIPE

Copper pipe is sized nominally according to its inside diameter. The outside diameter is always ⅛ inch more than the stated size. Actual inside diameter depends on the thickness of the walls. This varies by type. You have a choice of three types of copper water tube, as it's called, types *K, L* and *M*. Type *K* (heavy-wall, color code green) comes in both hard-temper (rigid and straight) and soft-temper (flexible). Diameters are ¼, ⅜, ½, ¾, and 1 inch.

Type *K* is for underground pipes such as a buried service line. Never bury copper pipe (or any pipe) in contact with cinders. Cinders eat 'em up.

Type *L* copper pipe (medium-wall, color code blue) also comes in both hard- and soft-temper and sizes from ¼ inch to 1 inch. It's used for interior plumbing and heating lines with solder, flare or compression fittings.

Type *M* (light-wall, color code red) copper pipe is the least costly because of its thin walls. Type *M* comes in ⅜-inch to 1-inch hard-temper only and must be used with soldered joints. Type *M* is suitable for interior plumbing or heating applications. It is your best bet, within the limitations.

Soft-temper pipe can be used for concealed piping. Hard-temper, since it makes a better looking job and is more resistant to denting, is often used for exposed piping.

Rigid copper comes in 1-foot lengths. Soft copper comes in coils of 30 or 60 feet. Soft-temper pipe is great for remodeling and modernization jobs where it can be worked down through existing partitions and walls through small openings.

Copper pipe is easiest to cut with a hacksaw in a miter box. Use a 24-tooth blade. A tubing cutter also may be used. After cutting, remove all burrs with a round file or reamer. Sizes to 1½ inch are easiest and best cut with a tubing cutter. Don't use cutting oil. If a hacksawed cut doesn't turn out square, file it square with a flat file.

The best way to hold a pipe while sawing it is in a miter box. This also ensures square cuts. A vise tends to crush it. If you must use a vise, clamp the pipe 6

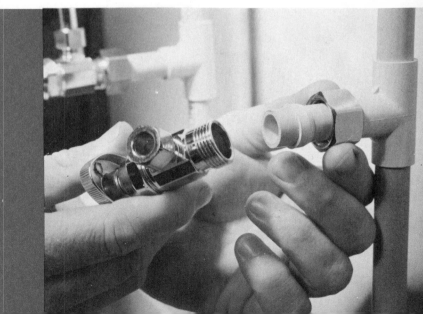

Use steel-to-plastic adapter to go from an existing steel pipeline to plastic system. From this point on, plastic pipe can run all the way.

The easy way to get from plastic water supply pipe to fixture: angle-stop is made to go with the pipe. Tighten nut, it's sealed.

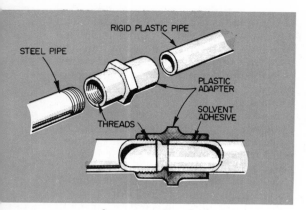

Adapter—steel to rigid plastic

inches from the end to leave the end uncrushed.

Fittings for copper water tube are of cast bronze, wrought copper and bronze, and brass. Wherever you have a choice of using a short-radius fitting or a long-radius one, choose the long radius. It passes water with less restriction.

SOLDERED JOINTS

Copper plumbing is most often done with soldered joints. These are sometimes termed *sweat joints*. The fittings for them cost the least of any.

Sweating a joint is easy. To do it, first polish the outside of the pipe and inside of the fitting with No. 00 steel wool or fine sandpaper. Get the metal bright and clean. Don't use a file. It scores the metal. The tube end must be round, not squashed. Cut off any out-of-round end and start again. Apply a thin coat of noncorrosive flux or soldering paste to the cleaned portions of the fitting and pipe. Slip the two together, removing any excess flux around the fitting. Heat the joint evenly with a propane torch or blowtorch. Move the torch back and forth to spread its heat.

Plastic pipe can be joined to flare fittings by flaring. Do this in usual way using a flaring tool, but be careful to avoid damaging edges. If they crack, cut and try again.

Test the temperature of the joint by touching it with solder as you heat. Test opposite the side being licked by the flame. When the solder will melt, remove the flame and feed solder into the joint. Capillary action will do the rest. Solder will even be drawn upward into a down-facing fitting. The joint is complete when a line of solder shows all around it. Remove any surplus solder by wiping or brushing. Let the joint cool before moving it. This shouldn't take more than a minute. Although 50/50 tin/lead solder is best, 40/60 solder may be used. Always use solid-core solder.

Sometimes to save time, all the fittings in one area can be cleaned, fluxed and assembled. Then all can be soldered in succession. Don't wait more than an hour to solder or the joints will oxidize again.

To avoid melting out a previously made solder joint when you sweat a new joint on the same fitting, wrap the completed joint with a wet rag.

Don't get a joint too hot or the flux will burn and the solder won't bond properly. An overheated joint must be taken apart, cleaned and resoldered. Likewise, any soldered joint that leaks after the water is turned on will have to be drained, taken apart, cleaned and resoldered.

If you ever have to unsolder a joint — and you might — heat it to melt the solder, then pull the pipe out of the fitting. Protect other connections in that fitting with wet rags.

When soldering a copper-to-copper valve, open it and wrap wet rags around the stem portion. This will prevent heat from damaging the valve's washer and packing.

Be careful to keep dirt and fingerprints off the prepared parts of a sweat joint. These will keep the solder from sticking to the joint.

Joints in pipes larger than 3/4 inch need to be heated on two sides for good heat distribution. Either use two torches or move the torch to heat on two sides.

Use care in working with a torch. It's fire. If a wood framing member is in the way of the flame, shield it with a piece

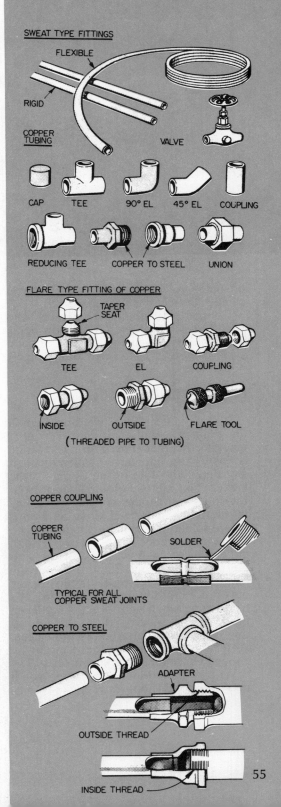

SWEAT TYPE FITTINGS

FLEXIBLE

RIGID

COPPER TUBING

VALVE

CAP   TEE   90° EL   45° EL   COUPLING

REDUCING TEE   COPPER TO STEEL   UNION

FLARE TYPE FITTING OF COPPER

TAPER SEAT

TEE   EL   COUPLING

INSIDE   OUTSIDE   FLARE TOOL

(THREADED PIPE TO TUBING)

COPPER COUPLING

COPPER TUBING

SOLDER

TYPICAL FOR ALL COPPER SWEAT JOINTS

COPPER TO STEEL

ADAPTER

OUTSIDE THREAD

INSIDE THREAD

55

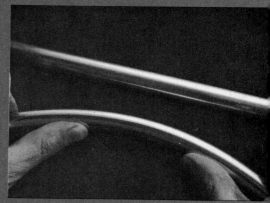

Copper pipe comes in three thicknesses, types K, L and M (from left to right above) and two tempers, one rigid and one flexible.

of metal. A cookie tray will do, if you can get it out of the kitchen unseen.

Copper pipe running through a stud wall must be protected from nails being accidentally driven into it. Put a metal strap over every stud across the pipe's notch-out.

FLARE FITTINGS

If you want an ideal working method for plumbing with soft-temper copper pipe, it's the use of flared fittings. Flaring is easy. Any line can be taken apart at any time. The cost of the fittings is the kicker. This method of joining is popular for splicing a copper house service line in the trench under wet, dirty conditions where soldering would be rough.

Flaring is done with a special tool, either the screw-down type or the simpler pound-in type. Be sure to put the flange nut over the pipe, threads facing outward, before you make the flare. It's maddening when you forget. The flange nut holds the pipe tightly to its flare fitting. Don't use flared joints in inaccessible locations where there might be enough vibration to work them loose.

With a pound-in flaring tool, you'll need a different tool for each pipe diam-

eter. Hold the end of the tubing in one hand while you tap the tool with a hammer in the other. Keep tapping until the tubing's end has been spread enough so that the threaded side of the flange nut will just fit over it. Take care to spread evenly all around.

A screw-in flaring tool has a series of various-sized openings between a pair of steel bars. Each opening is designed to take a single size of pipe. Find the right opening for the pipe you're flaring and clamp it tightly between bars with the chamfered edge of the bars toward the end of the pipe. The pipe should be flush with the bars. Then put the flaring cone over the tube and screw it down tight on the pipe's end. That does it.

With either flaring method, be sure there are no burrs or dirt to spoil the contact.

To assemble the flared joint, start the nut onto the fitting's threads by hand. Tighten as much as you can by hand making sure the threads are started right. Complete the joint by tightening securely with a pair of open-end wrenches, one on the flange nut, the other on the fitting. Never tighten the nut alone. It puts all the torque onto the pipes supporting your fitting. This can easily ruin the con-

To make a soldered joint in copper pipe you sandpaper end of pipe and inside of fitting. Apply flux sparingly and slip them together.

Heat fitting with propane torch. Move it around to spread heat evenly. Test temperature of the joint by touching with solder.

nection at the other side.

COMPRESSION FITTINGS

A third method of joining copper pipe is with a compression fitting. It consists of a fitting—coupling, angle, tee—plus a brass compression ring and flange nut for every pipe coming into it. The compression ring is slipped over the pipe after the flange nut. Then the nut is drawn down onto its fitting. This action squeezes the compression ring tightly between pipe and fitting, sealing the joint. No tools other than a pair of open-end wrenches are needed. Like flare fittings, compression fittings can be taken apart and reassembled whenever necessary. They're rarely used for extensive plumbing because of their cost. Solder fittings are much more economical. For an occasional small job, compression fittings are not bad.

Bending soft-temper copper pipe can be done in your hands. To bend rigid copper pipe you need a tubing bender to get a smooth job. A smooth bend is less restricting to the flow of water than an angle fitting. It also saves the cost of the fitting and making two soldered joints to go with it. Don't try to bend Type *M* pipe. The walls collapse.

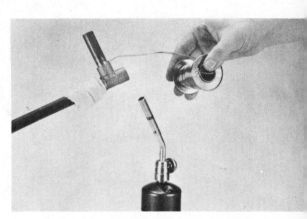

Take torch away and apply solder to heated fitting. It should flow in by capillary action, if temperature is right, and leave a fillet.

When you solder joints that are close to a flammable material, protect from heat with either a baking pan or other metal shield.

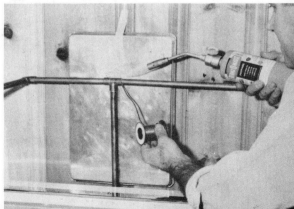

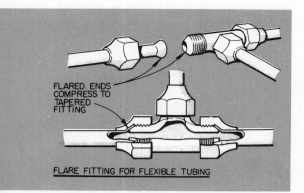

FLARED ENDS COMPRESS TO TAPERED FITTING

FLARE FITTING FOR FLEXIBLE TUBING

STEEL PIPE

Steel pipe comes galvanized (for plumbing) and in black iron (for gas lines). Its inside diameter is its size. For example ½-inch steel pipe measures ½ inch in inside diameter. You can buy 10-foot lengths, 21-foot lengths or lengths cut to order. If you buy cut-to-order lengths—as you should for small jobs— you may as well have them threaded too. For larger, more extensive work, you'll find it better to rent or buy a set of pipe

To flare a soft-temper copper tube with a screw-type flaring tool, clamp it in tool and tighten down on flare screw as far as possible.

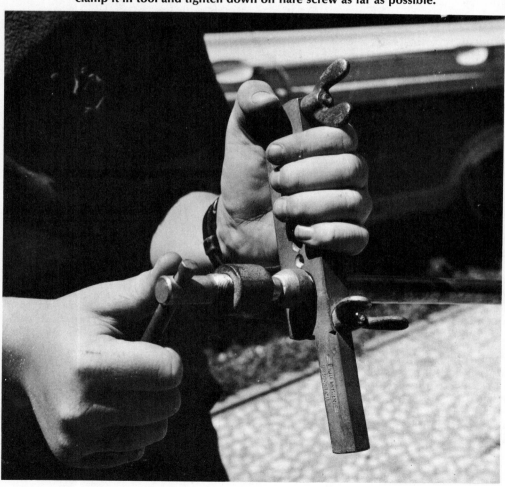

threading tools and have them on hand when you work.

Cut steel pipe with a hacksaw or pipe cutter. The cutter does the neatest job with the least effort. Lock a length of pipe to be cut in a pipe vise. This won't tend to crush it like a bench vise would. Many metal-working vises have a set of pipe jaws already in them. To use a pipe cutter, slip it over the pipe with the cutter wheel resting on the "cut" mark. Tighten the handle until the wheel bites. Apply some thread-cutting oil and rotate the cutter one turn. Then tighten again.

Keep rotating, oiling and tightening until the pipe has been severed. As you get almost through, support the pipe's end to keep it from sagging or falling as it is cut through.

To cut steel pipe with a hacksaw, take long, uniform strokes, applying pressure only on the "pull" strokes. The teeth, of course, should point toward the handle of the saw. Install the blade that way if it isn't already. File off any jagged edges on the outside of the pipe.

Burrs inside the pipe must be removed, otherwise they'll restrict the flow of wa-

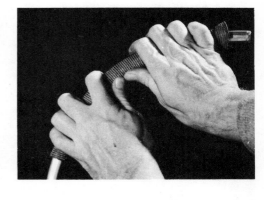

Smooth, kink-free bends in soft-temper copper pipe can be made using a tubing bender slipped over pipe to reinforce its walls.

The flare nut, which was inserted before making flare, will hold pipe in tight seal with any flare fitting. Fittings are costly.

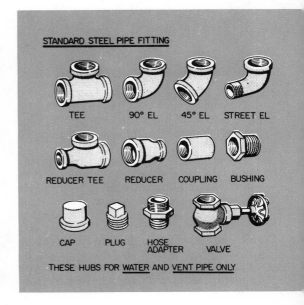

STANDARD STEEL PIPE FITTING

TEE   90° EL   45° EL   STREET EL

REDUCER TEE   REDUCER   COUPLING   BUSHING

CAP   PLUG   HOSE ADAPTER   VALVE

THESE HUBS FOR WATER AND VENT PIPE ONLY

To make threaded joints in steel pipe, cut pipe to the proper length with pipe cutter. With each cutter rotation, tighten handle.

Ream out burr left on inside of cut pipe. Use pipe reamer or, lacking reamer, use a rounded file. A pipe vice and stand help.

Thread pipe with stock and die. Turn stock clockwise. Back off quarter-turn, if it is hanging up, to clear chips. Also use oil.

Apply joint compound in pipe's threads and assemble the joint. Make sure the threads start straight. Tighten with pipe wrenches.

ter. Use a rounded file or a tapered pipe reamer chucked in an electric drill or bit brace.

Steel pipe is threaded with a pipe die held in a *stock* (handled die holder). The stock has a square opening on one side for insertion of the square die. On the other side is a pipe guide. The guide gets the die started true. Use the same size die and guide as the pipe to be threaded.

The printed side of the die should be placed facing away from the guide. If you put it in backward you'd have real trouble starting the die and the threads

wouldn't come out tapered, as they should.

Place the stock over the end of the pipe being threaded. As the die contacts, begin rotating the stock handle clockwise. Keep pushing while getting the threads going. Apply thread-cutting oil generously to the pipe and die. Once started, you can stop pushing. The threads will take hold and pull. Rotate slowly and continuously until the pipe's end sticks out one full thread beyond the outer die face. Then back off the die by rotating counter-clockwise. If the die

To make a holder for sawing pipe, clamp a pair of 2x4's to benchtop with space between to place pipe. Pipe is not damaged at all.

should tend to hang up while threading or backing off, unhang it by reversing the rotation for a quarter-turn. This should clear any cuttings out of the way. They're usually the hang-up. Clean up excess oil and chips with a cloth.

Before making up a steel pipe joint, remove dirt and chips from inside the pipe and around the threads. Apply pipe joint compound to the outside threads only. Never to the inside threads of the fitting. Use pipe joint compound sparingly with just enough to fill the threads evenly. No excess or barren spots. Don't get compound inside the pipe or over its end. Start the threads by hand to make sure they're not cross-threaded.

Steel pipe fittings are tightened with a pair of pipe wrenches, one on the pipe and a larger one on the fitting. A 10-inch wrench will handle pipes to 1 inch, an 18-inch wrench, to 2 inches.

A word of warning. While a threaded joint must be leakproof, you can easily overtighten the threads of ½-inch or ¾-inch pipes. With a new fitting and newly threaded pipe made as described, the joint will be tight enough when about three threads are still visible outside the fitting. Each time you remake the same joint, however, you'll have to tighten a little more to get the proper fit. Never, never turn a pipe in more than one turn

after the last thread has disappeared inside the fitting. You'll likely blow the fitting if you do. After you've made up a few joints, you'll quickly get the feel of how tight a joint should be.

STAINLESS STEEL PIPE

A new kind of water supply pipe now on the market is made of stainless steel. It's called *Ti-Krome* by the manufacturer, Tubotron, Inc. The cost is about one-third that of equivalent grade copper pipe. Regular copper sweat fittings should be used until stainless steel fittings are made available. The procedure is much the same as for joining copper water tube, except that a corrosive flux is used. This must be thoroughly cleaned off before you leave the joint.

As you might expect, stainless steel pipe is highly resistant to corrosion and isn't much affected by mineral deposits from hard water. The pipe can be cut with a hacksaw or tubing cutter, soldered with your propane torch. You can bend it with a pipe bender the same as rigid copper tubing but it's tougher to bend because of its strength. Stainless steel pipe is too new to be included in many building codes, so you may have to interpret code provisions regarding its use. The pipe should be available through your local plumbing supply dealer.

DWV PIPES AND FITTINGS

Drainage pipes must be large and slope toward the point of disposal

There are enough drain-waste-vent piping materials to suit every code and every plumber. PVC and ABS plastic pipe are ideal for the home handyman. You'll also like running copper DWV lines. If you're a cast iron pipe fan, then you'll want to consider no-hub, the modern method of joining it. With cast iron pipe, threaded galvanized steel pipe is usually used for the smaller waste and vent lines.

All drainage fittings must be designed for that purpose. They're called *sanitary* fittings. Sanitary fittings provide for a free flow of water. They have no ridges or pockets to collect solids. A drainage fitting should always be installed facing in the proper direction. Vent runs, since no water passes through them, can use ordinary (not sanitary) fittings. Plastic and copper fittings in the DWV sizes all are sanitary types. When used for vent runs they should be inverted to pass the vent gases up, not down.

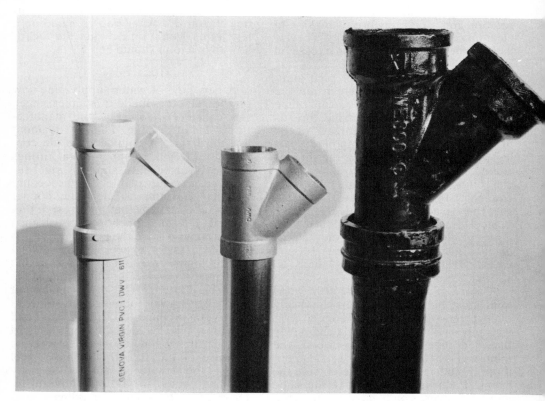

Take your pick from among these drain-waste-vent pipe types. They are, left to right, 3-inch plastic, 3-inch copper and 4-inch cast iron.

Plastic pipe for DWV is easy to install using the solvent welded joints described for water supply pipes. They are done the same, except a larger brush is needed for the larger pipe sizes. PVC plastic pipe can be leaded into a cast iron fitting using the usual technique. No special fitting is needed if the sizes match. If they don't, a transition fitting is required.

For its PVC pipe Genova Products has worked out an indexing system. Numbers appearing on the fittings indicate angles of 45° and 90° all the way around. When the proper number is lined up with a black plumbline on the pipe, perfect alignment is assured. Changes in alignment are made just as easily. The reference marks permit fast work without directional errors. It's great.

Plastic DWV pipe comes in 1½-, 2- and 3-inch diameters, 10-foot lengths. Sometimes 2-inch is hard to find. A 3-inch plastic pipe will fit in a 2×4 stud wall, a distinct advantage of plastic over cast iron pipe. A disadvantage: if you goof up on a fitting's placement or alignment, kiss it goodbye.

As the manufacturers of other types of pipe point out, plastic pipe is weakened or dissolved by powerful solvents, such as acetone. If you're going to be pouring acetone down your drains, don't use plastic. But in ordinary household use, there's no problem. PVC plastic is safe for use in the presence of acetic, nitric, hydrochloric and sulfuric acids as well as ammonium hydroxide, sodium hydroxide, sodium chloride (common salt), petroleum oils, etc. It is likewise resistant to rodent, termite and bacteria attack. ABS does about as well.

Copper drain-waste-vent piping comes in 10- and 20-foot lengths. You can also buy it cut to-length by the foot. Like plastic, its long lengths keep joints to a minimum. Copper is light, easy to cut and pleasant to work with. The solder-type fittings are trim and compact. The largest size you need is 3-inch. Thus a copper soil stack fits within a standard 2×4 partition. Copper resists corrosion. The joints are strong and are simply made with a blowtorch or pair of propane torches, flux and wire solder. Copper fit-

D-W-V VALUE ENGINEERING

3" COPPER

3" PLASTIC

4" CAST IRON

3" NO-HUB CAST IRON

For ease of installation, plastic DWV system can't be beat. Solvent-welded joints set firmly, seldom give trouble.

RECESSED THREAD ALLOWS FREE FLOW

PIPE END WOULD SNAG SOLID WASTES AND CAUSE CLOGS

DRAINAGE FITTING

VENT OR WATER ONLY

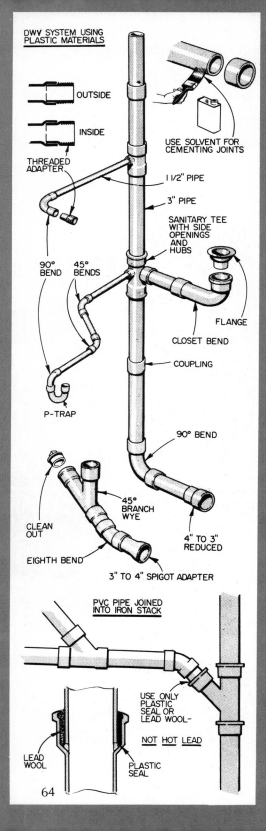

DWV SYSTEM USING PLASTIC MATERIALS

OUTSIDE

INSIDE

THREADED ADAPTER

USE SOLVENT FOR CEMENTING JOINTS

1 1/2" PIPE

3" PIPE

SANITARY TEE WITH SIDE OPENINGS AND HUBS

90° BEND

45° BENDS

FLANGE

CLOSET BEND

COUPLING

P-TRAP

90° BEND

45° BRANCH WYE

CLEAN OUT

4" TO 3" REDUCED

EIGHTH BEND

3" TO 4" SPIGOT ADAPTER

PVC PIPE JOINED INTO IRON STACK

USE ONLY PLASTIC SEAL OR LEAD WOOL—

NOT HOT LEAD

LEAD WOOL

PLASTIC SEAL

64

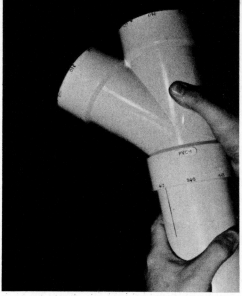

Angle markings on some DWV fittings let you adjust accurately the alignment of a joint with line on pipe before they set.

tings in DWV sizes are costly, you'll find. Don't use drainage-type copper below ground.

Copper piping for drain-waste-vent use comes in two weights: DWV and Type *M*. DWV is thinner-walled and about 40 percent cheaper. Use it if your code permits. Available is 1½-, 2- and 3-inch copper pipe, all with sanitary fittings to match.

CAST IRON

There's little doubt about it. Once a cast iron soil pipe system is in, it's the best material. Getting it in, that's the problem.

Walls containing cast iron piping can be nailed into without sweat. This is a help in picture-hanging as well as in applying wall materials. Corrosion is no problem with cast iron. Moreover, cast iron pipe will withstand stresses that would break up other types of piping.

Ordinary cast iron pipe comes in lighter *service weight*, and *heavy weight*. Code permitting, use service weight. The swingingest way to use cast iron soil pipe is with the great no-hub joints. They're

Transition fitting joins a 3″ copper pipe to 4″ cast iron. Fitting is sweated to copper, joined with lead and oakum to iron.

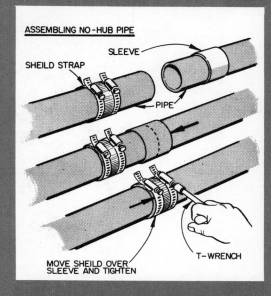

SHEILD STRAP

SLEEVE

PIPE

MOVE SHEILD OVER SLEEVE AND TIGHTEN

T-WRENCH

tops for compact installations in 2×4 walls, the kind you'll be doing in home remodeling. New no-hub fits in beautifully with existing hub-type cast iron piping, too.

The no-hub joint consists of special hubless pipe and fittings, a neoprene gasket and stainless steel sleeve and clamps that go over the gasket. No-hub installs fast, clean and at low cost. The rubber gasket acts as a cushioning element to take up shocks and vibrations within the drainage system. It takes the "gurgle" out of plumbing.

The Cast Iron Soil Pipe Institute developed no-hub and it's produced in one weight only. No-hub is designed around a minimum number of fittings to keep things simple. Unlike hubbed cast iron pipes, there are no waste pieces left over. Cut-offs are usable, every time. All fittings are the sanitary type. These can be installed upside down for vent runs. Any accessible pipe or fitting can be removed for repair or alteration of the line without disturbing the adjacent fittings. No-hub can be used above ground or below (depends on code). A 3-inch no-hub line fits into a 2×4 wall without furring it out.

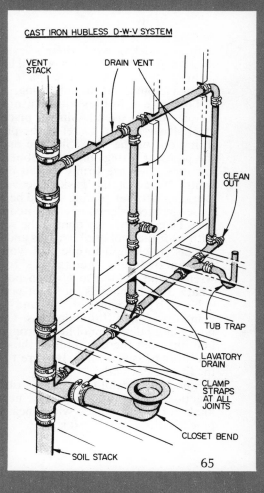

CAST IRON HUBLESS D-W-V SYSTEM

VENT STACK

DRAIN VENT

CLEAN OUT

TUB TRAP

LAVATORY DRAIN

CLAMP STRAPS AT ALL JOINTS

CLOSET BEND

SOIL STACK

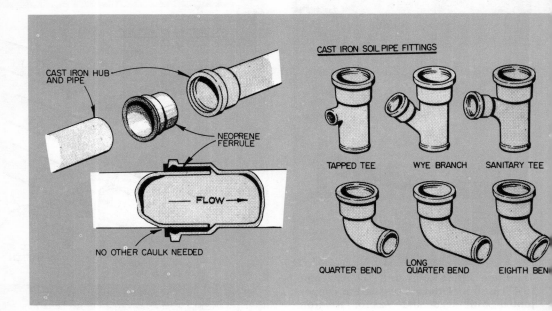

CAST IRON SOIL PIPE FITTINGS

CAST IRON HUB AND PIPE

NEOPRENE FERRULE

← FLOW →

NO OTHER CAULK NEEDED

TAPPED TEE   WYE BRANCH   SANITARY TEE

QUARTER BEND   LONG QUARTER BEND   EIGHTH BEND

In remodeling work there's less wall to cut and no need to burn out old leaded joints. Merely cut out a space large enough to put in the no-hub fitting.

To make a hubless joint a no-hub torque tool is useful. Simply place the steel sleeve on the barrel of the pipe or fitting out of the way. Slip the ends of the pipes or fittings into the rubber gasket until they are firmly butted on the center flange of the gasket sleeve. No lubricant or other coating should be used. Slide the steel sleeve over the neoprene one. Then alternately tighten the screws with a torque tool until it clicks and freewheels. Lacking a torque tool, you can use an ordinary mechanic's torque wrench and tighten to 60-inch pounds. That's INCH pounds, not foot pounds.

No-hub pipe comes in 2-, 3- and 4-inch diameters, 5- and 10-foot lengths. There's no required direction of flow through no-hub pipe, but there is through the drainage fittings used with it. Be sure to face the fittings correctly.

Since the ends of the pipes (or fittings) practically touch, no allowance need be made for joining space when measuring pipe lengths. Cut it like it is.

To calk a soil pipe joint, assemble the spigot in hub and ram rope oakum around the joint with a plumber's calking tool.

66

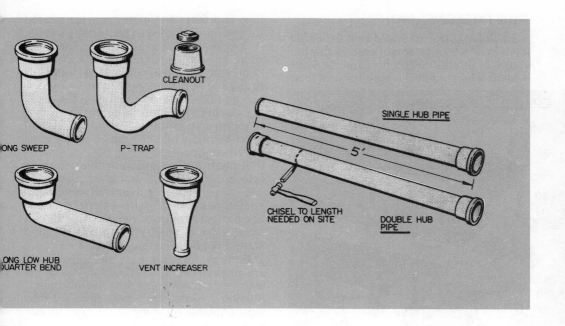

LONG SWEEP

P-TRAP

CLEANOUT

LONG LOW HUB
QUARTER BEND

VENT INCREASER

SINGLE HUB PIPE

5'

CHISEL TO LENGTH
NEEDED ON SITE

DOUBLE HUB
PIPE

For horizontal joints install a joint runner and pour lead into joint at the top. Melt enough to make one continuous pour.

Use calking tool to pack lead against the spigot and hub surfaces. Go around joint three times; lead depth is ¾ to one inch.

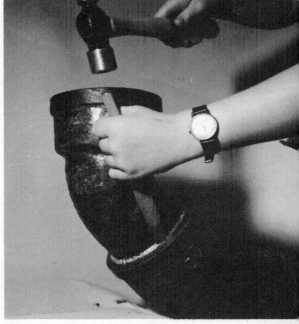

HUB-TYPE CAST IRON

Each length of standard cast iron pipe has a hub end and a spigot end. In use the spigot is fitted into the hub and the joint sealed. The standard length for cast iron pipe is 5 feet. A 4-inch diameter is used for main soil stacks, those serving toilets. A 2-inch diameter also is available for running smaller drains and secondary vent stacks, those without a toilet attached. Cast iron pipe comes in single- and double-hub. Double-hub is used for cutting. The cut-off piece has a hub of its own and thus is not wasted. A wall must be made of 2×6 studs to carry runs of 4-inch pipe.

Branch drains and vent lines running into a cast iron soil stack are usually of 1½-inch threaded galvanized steel pipe, sometimes 2-inch, with cast iron drainage or vent fittings. Secondary stacks can usually be of steel pipe too. Steel pipe should not be used underground.

Fittings for cast iron soil pipe itself are made in every combination you could want for running soil stacks and drains. Smaller 1½-inch or 2-inch threaded side taps available on some fittings permit entrance of waste and vent lines.

Service weight cast iron pipe is cut by hacksawing around it to a 1/16-inch depth. Get the groove square around the pipe. Then lay the pipe over a 2×4 with the groove at the edge of the board and tap on the pipe's end with a heavy hammer. Pound progressively harder until the end breaks off around the cut. Use chalk to mark for cutting.

COMPRESSION JOINTS

The compression joint with rubber flange gasket is the modern method for joining hub-type cast iron soil pipe. By all means use it if you can (and if you don't go the no-hub route).

The compression joint uses a one-piece neoprene rubber gasket that you insert in the hub portion of the fitting. A beadless spigot (cut the bead off if you can't get it that way) is inserted into the lubricated gasket. Jolting of the pipe on the floor, or leverage forces the spigot into the gasketed hub. The joint is sealed by displacement and compression of its gasket. Be sure the parts to be joined are clean and free of burrs and metal fins.

A compression connection is leak-proof, root-proof and pressure-proof. It

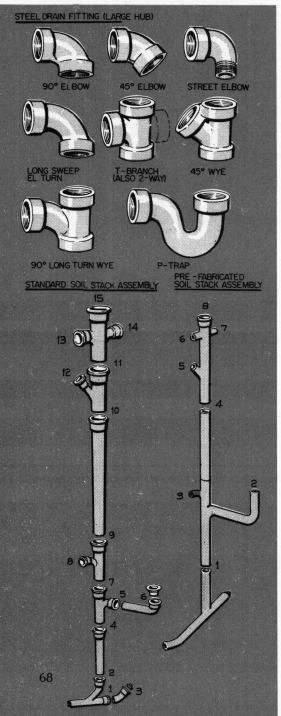

STEEL DRAIN FITTING (LARGE HUB)

90° EL BOW   45° ELBOW   STREET ELBOW

LONG SWEEP EL TURN   T-BRANCH (ALSO 2-WAY)   45° WYE

90° LONG TURN WYE   P-TRAP

STANDARD SOIL STACK ASSEMBLY   PRE-FABRICATED SOIL STACK ASSEMBLY

absorbs vibration and deflection of up to 5 degrees without leakage. This method is now widely accepted.

LEAD AND OAKUM

If you must do things the hard way, use the old standby calked joint. If you do, the apparent economy of cast iron is shot down somewhat by costs for lead.

Place vertical runs of pipe with the hub end of each pipe and fitting upward. The flow must always be in at the hub end and out through the spigot end. See that the ends of the hubs and spigots are clean and dry. Place the spigot in the hub all the way, centered and plumb. Support it in position. Make the following length allowances for depth: 2-inch pipe, 2½ inches; 3-inch pipe, 2¾ inches; 4-inch pipe, 3 inches. Measure pipe from the inside shoulder of the hub out to the distance required.

Melt enough lead to fill the joint in one pour. A plumber's furnace is a big help if you have many joints to do. A 1-inch deep joint uses one pound of lead per inch of pipe diameter. If your code allows only a ¾-inch depth of lead, melt ¾ pound per inch of pipe diameter.

While your lead is melting, pack rope oakum into the joint. You'll need an ounce of it per inch of pipe diameter. Wrap the oakum around the pipe and push it down into the hub with a plumber's calking tool. Keep packing oakum in until the joint is filled to the proper depth. Pack hard. Pound. The oakum should make a watertight seal. Don't count on the lead alone to do it.

The best way to handle molten lead is in a dry ladle. Lead should be heated a few minutes beyond melting but not too long. If too hot, it will burn the oakum. Pour lead into the joint until it's filled to brimming. When the lead cools and shrinks, ram it down by pounding the calking tool into it. Work the lead tightly against both pipe and hub all the way around. Go around several times to make sure.

To calk a horizontal joint, you'll need an asbestos joint runner to keep the lead from leaking out while you pour. Place the runner's clamp at the top to form a funnel for pouring. Tap the joint runner into tight contact with the hub for a leak-proof seal. Make the pour all at one time. Fill to overflowing. When the lead has solidified, you can remove the runner and pack the lead with your calking tool.

Cut off any excess lead with a cold chisel. A horizontal run with several joints is easier to make if you stand the pipes vertically to join them. Then install the prefabricated length.

Lead wool is easier to use than molten lead, but costs more. To use it, twist it into a rope, like oakum, lay it in the joint and calk it just as you would poured lead. The strands should be half again as thick as the space to be filled and uniform throughout. There should be no thin or lumpy spots. Calking will not bond the strands unless they're twisted together. Cold lead can't be joined to cold lead by hammering.

THREADED SYSTEM

Instead of cast iron you can use what's called a *Durham* system. All joints are threaded, soil stack and all. Durham fittings are the threaded cast iron sanitary fittings already described for waste and vent lines. The pipe is regular threaded galvanized pipe in diameters to 3- and 4-inch. Measuring, cutting and threading on big pipes is no snap. Use this system only if you like that kind of work.

You can sometimes buy soil stack parts prefabricated so that four pieces with three joints will take the place of some 9 pieces with 8 joints. Prefabs are made of cast iron and take leaded joints.

If you build your own plastic or copper system, you can save work by prefabricating parts of it. Joints can be assembled on the workbench and the larger units erected into place. Don't go too far and work yourself into a corner, however.

Some codes call for the use of a lead connector between the toilet and soil stack piping. The idea is to permit some movement between the two. Either make use of standard lead connections (ask your dealer) or call in a plumber to cut and make a *wiped* joint at the toilet flange. It's not easy to do yourself.

When soldering, wrap completed joint with wet cloth so the old solder does not melt.

TIPS ON RUNNING PIPES

Plan water lines to travel the most direct routes to fixtures

Modern fittings and fixtures have taken many problems out of running pipes. You don't need to be an experienced plumber to do it. Anyone who can follow a few simple directions can turn out a creditable job the very first time. It's merely a matter of measuring, cutting and assembling the pipes and fittings according to a set of sensible rules.

Pipes need carry only so much water. Beyond a certain point a larger pipe size is only wasteful. The following are fairly standard sizes for house pipes.

Service entrance — 1 inch.

Service to water heater — ¾ to 1 inch.
Hot and cold mains — ¾ inch.
Branches to sinks, showers, bathtubs, laundries and dishwashers — ½ inch.
Branches to lavatories and toilets — ⅜ inch.
Soil stacks and house drains — 3 to 4 inch.
Sewer lines — 4 inch.
Branch drains from tubs, showers, sinks, laundries, lavatories, dishwashers — 1½ inch.
Roof vents — 3 inch or more.
Many of the newer lavatory faucets

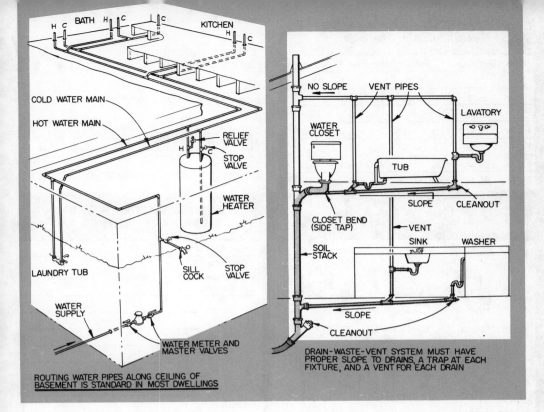

BATH   KITCHEN

COLD WATER MAIN
HOT WATER MAIN
RELIEF VALVE
STOP VALVE
WATER HEATER
LAUNDRY TUB
SILL COCK
STOP VALVE
WATER SUPPLY
WATER METER AND MASTER VALVES

ROUTING WATER PIPES ALONG CEILING OF BASEMENT IS STANDARD IN MOST DWELLINGS

NO SLOPE   VENT PIPES
WATER CLOSET
LAVATORY
TUB
SLOPE   CLEANOUT
CLOSET BEND (SIDE TAP)   VENT
SOIL STACK
SINK   WASHER
SLOPE
CLEANOUT

DRAIN-WASTE-VENT SYSTEM MUST HAVE PROPER SLOPE TO DRAINS, A TRAP AT EACH FIXTURE, AND A VENT FOR EACH DRAIN

come equipped with ¼-inch supply pipes from the main branch to them. If you run ⅜-inch pipes up to these, you get shot down right at the faucet. Code permitting, you can probably get along with ¼-inch branches to these lavatories.

HOW LONG A PIPE?

Measuring for pipes and cutting them would be a snap except for two things: *fitting gain* and *make-up*. Fitting gain is the amount of space taken up by the fitting a pipe goes into. Pipes in a run, or ones coming to an angle, don't butt up against each other inside the fitting. The fitting separates them somewhat. Make-up is the amount that a pipe goes into the fitting. Both fitting gain and make-up must be allowed for in measurement and cutting of pipes.

Threaded steel pipe fittings are so standardized that the gain and make-up for them can be given in a table. There's one table for standard fittings another for drainage fittings. Plastic and copper

water supply fittings and most drainage-waste-vent fittings vary in design with make. Take actual measurements of the fittings you will use and then determine gain and make-up.

Measurements for steel pipe are usually taken in one of three ways: end-to-end for a single pipe or for a pipe run with fittings; end of pipe to center of fitting for angled runs; or center of fitting to center of fitting for parallel runs.

Instead of holding a steel fitting up while taking your measurement, use the

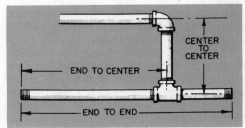

CENTER TO CENTER
END TO CENTER
END TO END

Three ways of measuring pipes.

71

ALLOWANCES FOR THREADED STANDARD FITTINGS

Pipe Size	Distance "X"	A	B	C	J	K
½ in.	½ in.	1⅛	⅞	1⅝	1⁵⁄₁₆	1¼
¾ in.	½ in.	1⁵⁄₁₆	1	1⅞	1½	1⁷⁄₁₆
1 in.	⁹⁄₁₆ in.	1½	1⅛	2⅛	1¹¹⁄₁₆	1¹¹⁄₁₆
1¼ in.	⅝ in.	1¾	1⁵⁄₁₆	2⁷⁄₁₆	1¹⁵⁄₁₆	2¹⁄₁₆
1½ in.	⅝ in.	1¹⁵⁄₁₆	1⁷⁄₁₆	2¹¹⁄₁₆	2⅛	2⁵⁄₁₆
2 in.	¹¹⁄₁₆ in.	2¼	1¹¹⁄₁₆	3¼	2½	2¹³⁄₁₆

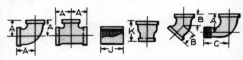

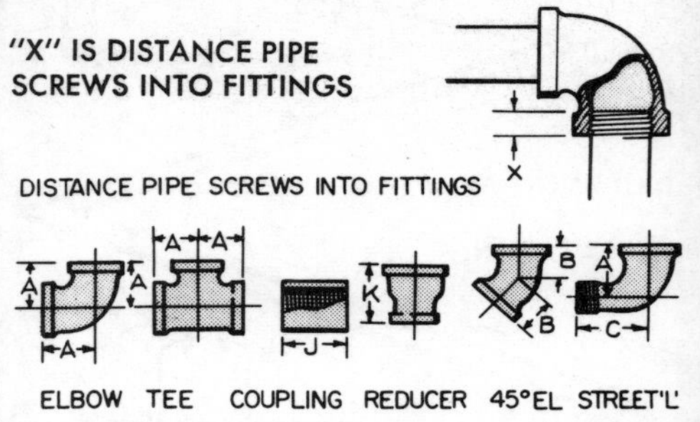

"X" IS DISTANCE PIPE SCREWS INTO FITTINGS

DISTANCE PIPE SCREWS INTO FITTINGS

ELBOW   TEE   COUPLING   REDUCER   45°EL   STREET'L'

ALLOWANCES FOR THREADED DRAINAGE FITTINGS

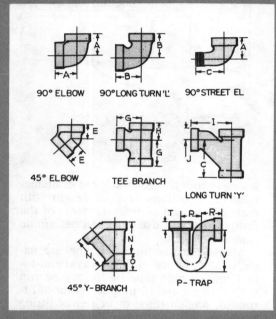

90° ELBOW   90° LONG TURN 'L'   90° STREET EL

45° ELBOW   TEE BRANCH   LONG TURN 'Y'

45° Y-BRANCH   P-TRAP

Pipe Size (in.)		1½	2
Distance "X" Pipe Screws into Fittings		⅝ (inches)	¹¹⁄₁₆ (inches)
	A	2³⁄₁₆	2⅜
	B	2½	3¹⁄₁₆
	C	2¹¹⁄₁₆	3¼
	E	1⁷⁄₁₆	1¾
	G	2½	3¹⁄₁₆
Fitting Dimensions (inches)	H	1¾	2⅛
	I	4⅛	5⁷⁄₁₆
	J	1¼	1⅝
	K	4⅛	5⁷⁄₁₆
	N	3⅝	4⁵⁄₁₆
	O	1⅞	2⅛
	P	2¼	2¾
	R	2⅛	2⁹⁄₁₆
	T	⅞	⅞
	V	5	5¹¹⁄₁₆

table to see how much space the fitting will occupy. Then subtract this amount from your measurement. Finally add on distance "X" and you have it.

Here's an example. A length of ½-inch pipe is to be cut running between two other pipes (see drawing). One pipe has a tee, the other is to be fitted with an elbow. Suppose you measure 29 inches center to center of the existing pipes. For both the tee and elbow the fitting gain from the table is 1⅛ inch. Make-up for

½-inch pipe is ½ inch. Subtracting the 1⅛-inch fitting gain at each end gives 26¾ inches. Adding the 1-inch make-up loss gives 27¾ inches. This is the actual length the pipe should be cut. Measure the pipe and mark it with a file or hacksaw.

FACE-TO-FACE METHOD

A simpler face-to-face method for pipe

measurement requires having the fittings in position. To find the length of a run of the pipe between two fittings, measure the face-to-face distance between those fittings. Then add twice the make-up measurement for that pipe size from the table (distance "X"). The make-up measurement is doubled because the pipe screws into fittings at both ends. Measured and cut to that total length, the pipe should fit perfectly when tightened in place.

Suppose in the previous example that both fittings were installed and you found a 26¾-inch face-to-face distance between them. Adding on twice the ½-inch make-up points to the use of a 27¾-inch length of pipe. Simple!

Even steel pipe expands as it's heated. For this reason all pipes — whether steel, copper or plastic — should be fastened to framing with pipe hangers so that some expansion can take place. Extra long runs of copper or plastic water supply pipe need an expansion loop or a special copper expansion fitting. An expansion loop is a "U" in the line. One is made with four 90-degree elbows and short lengths of pipe. Steel and rigid copper water supply pipes should be supported every 7 to 10 feet. Soft-temper copper water supply pipes need support every 16 inches. Vertical runs of pipe between the floors of a house can rest at the lower end on a header nailed between floor joists. DWV pipes need support every 4 feet (every third joist), except no-hub, which should be supported at every joist on horizontal runs.

All water supply piping should be pitched slightly backward from the fixtures to a stop-and-drain valve placed at the low point. Often this is located at the water meter. If there are other low points in the system, these should have provision for drainage too. The drain side of a stop-and-drain valve is always placed where it will not be under pressure when the valve is off. These valves permit draining the system if the house is not to be heated during below-freezing weather.

Because there is no water pressure to move the flow along in drain-waste-vent piping, all horizontal DWV runs are not

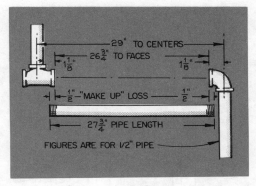

Figure fitting-gain and make-up loss.

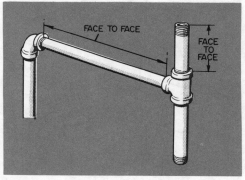

Take actual measurements and add make-up distances for pipe size, to measure pipe.

Use compact drill to bore holes for pipe runs, because other electric drills can't fit.
Stanley Tools

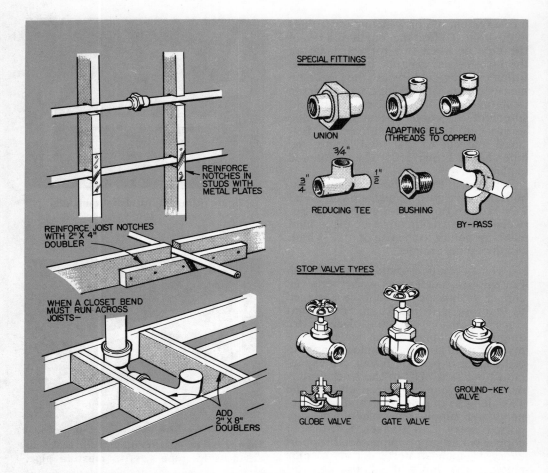

REINFORCE
NOTCHES IN
STUDS WITH
METAL PLATES

REINFORCE JOIST NOTCHES
WITH 2" X 4"
DOUBLER

WHEN A CLOSET BEND
MUST RUN ACROSS
JOISTS—

ADD
2" X 8"
DOUBLERS

SPECIAL FITTINGS

UNION

ADAPTING ELS
(THREADS TO COPPER)

3/4"

3/4

1/2

REDUCING TEE   BUSHING

BY-PASS

STOP VALVE TYPES

GLOBE VALVE   GATE VALVE

GROUND-KEY
VALVE

strictly horizontal. Instead they should slope ¼ to ½ inch per foot.

PIPE RUNS

With threaded steel pipe, begin each run where it originates, such as at the water heater or softener. Take it to its end. This will eliminate unions in the line. They're costly and bothersome to install. With plastic or copper pipe use any installation sequence that seems practical. You won't need unions unless you want them at appliances. Information on installing DWV piping is given in the chapter on a home addition.

Pipes in a basement or crawlspace are usually run beneath the joists and fastened at intervals with pipe hangers. If they are run up between joists, use a pair

of 45-degree elbows (or ⅛-bends) to make the transition from above or below to between. Two 90-degree elbows can be used too in water supply lines, but they are more restrictive to water flow than a pair of 45's. Attic runs are generally made above the joists. Hangers are not needed there.

In new construction, pipes can be run under floors, over ceilings and inside walls. For a remodeling job you may have to fur out an existing wall, or box in a soil stack to conceal it. Pipes than run across a partition have to be notched into the studs. Since the largest of these pipes is usually 1½ inches, there's no real problem. To avoid weakening studs, don't notch them any deeper than necessary. Then after the pipe is in place, cover each notch with a strip of steel.

Pipes running across floor joists are more of a problem. Notching joists is liable to weaken them. Never notch a joist more than one quarter of its depth. Never notch near the ends and never in the center half. If you must break these rules, reinforce the dishonored joist by spiking a 2×4 or 2×6 on one or both sides at the notch. A hole can be drilled anywhere in a joist if it's roughly centered between the top and bottom edge.

The best way to run hot and cold water supply pipes is side by side. Keep them 4 to 6 inches apart to prevent the hot one from radiating its heat to the cold one. Side by side makes them easier to install.

If the space where pipes run is unheated, the pipes should be insulated to keep water in them from freezing. It may be a good idea to insulate hot water pipes anyway, to prevent heat loss. Pipes between joists can be protected by stretching blanket insulation between the joists. Pipes across joists are harder to insulate. They have to be wrapped with strips of insulation.

SPECIAL FITTINGS

In addition to the more or less standard pipe fittings shown in the chapter on pipes and fittings, there are special ones you should know about.

Because steel threaded pipe cannot be turned to screw into a plumbed-in fitting at both ends, a union is required at some point in the run. A union is a coupling that comes apart without disassembling the pipe run it's in. It joins two pipes of the same size and type.

To join pipes that are the same type but not the same size, use a *reducer*. There are reducing couplings, reducing elbows and reducing tees. To reduce a threaded opening in a fitting, use a reducing bushing.

When the pipes are the same size, but not the same type, an *adapter* is used. There are adapter couplings, adapter elbows and adapter tees. Seldom used are very specialized reducing-adapting fittings that combine both features.

Using special fittings, plus the stan-

Make copper drain-waste-vent watertight joints by using big-flame tip on the torch.

If two dissimilar metals are joined together, like plastic to copper, use lots of pipe dope.

dard ones, you can assemble any pipe run even though it may change size and type.

You'll have the most need for two types of valves in home plumbing—the gate valve and the globe valve. Here is the rule for which type to use: If "off" or "on" is all you want, use a gate valve. It doesn't restrict the flow of water through it. If you want to control the flow of water, use a globe valve or other flow-control valve.

HOW TO HOOK UP FIXTURES

Fixtures, except the bathtub, are installed after walls are finished

Installing plumbing fixtures is completely different from putting in piping. Fixture installation is called *finish plumbing*. You have to do a finished looking job. That's not so tough though. The hard part is that what you learned about piping for water supply and DWV doesn't apply to finish plumbing. You have a different set of pipes and fittings to work with. It's a whole new ball game.

Take water supply hook-up. Every fixture needs pipes leading from each stub-out on the wall to the fixture tailpiece. The stub-outs may be ½- or ⅜ inch threaded steel pipe, copper pipe, or plastic pipe. The faucet itself comes in differ-

ent types too. There's the straight threaded-end kitchen sink faucet with coupling nuts. Some of these also come with short brass tailpipes having the bayonet tapers at one end and pipe threads at the other. Still other faucets come with flexible copper water tubes. Further complicating supply hook-up to a fixture is the fact that your water supply stub-outs may come either out from the wall or up through the floor.

The bright spot in the whole picture is a great product, the flexible fixture supply pipe. Plumbers call them *speedees*. A speedee is a bendable length of chrome-plated copper tubing in ⅜-inch diameter and lengths sufficient to reach between any stub-out and the faucet tailpiece. Speedees come with bayonet ends, flanged ends or screw ends. While a speedee costs more than building some sort of pipe hook-up to supply the fixture, the time and effort saved are well worth it.

ATTACHING TO STUB-OUTS

One end of each speedee has to be fitted to your pipe stub-outs. Here are your choices of fittings:

Fixture angle-stop — For wall stub-outs your best bet is to use chrome-plated fixture angle-stops. They're widely available. One end threads onto the male threads of ⅜- or ½-inch stub-outs. The other end has a ⅜-inch compression fitting to accommodate the speedee. The stop valve in between provides fast shut-off, for fixing a leaky faucet or emergency use, simply by reaching underneath the fixture.

To use an angle-stop, an escutcheon plate should be first slipped over the pipe stub with a ring of plumber's putty behind it to hold it to the wall. The stub-out should stick out 1 inch from the wall. Screw the angle-stop onto the pipe stub and align it facing up toward the fixture.

An angle-stop valve hooked to the wall stub-out helps in connecting the flexible supply pipe to the house plumbing, and furthermore, it provides a shut-off.

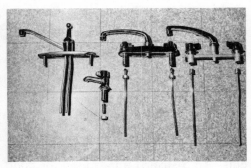

Types of faucet supply fittings are, left to right, ⅜″ soft copper tubes; threaded tailpiece, lavatory type, which takes bayonet-end speedees; finally, the sink type, which takes threaded-end speedees.

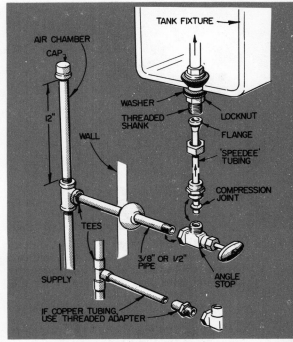

Wall fixture supply hook-up.

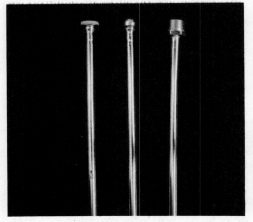

Fixture supply pipes (speedees) come with three different ends, from left to right: flat flange for toilet tanks; bayonet end for lavatories; and threaded for sinks.

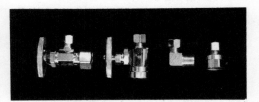

Transition fittings between stub-outs and fixture supply pipes are, from left to right: straight-coupling fitting, angle coupling, straight-stop and angle-stop.

The speedee is then cut to length and bent to fit between the angle-stop and the faucet tailpiece. A flange nut and compression ring secure the joint between the speedee and the angle-stop.

To use an angle-stop with copper pipe stubs, saw them off and fit them with male threaded adapters sticking out 1 inch from the finished wall. Special angle-stops are made for use with ½-inch plastic pipe stubs. To install one of these, put on the escutcheon, then slip the large flange nut onto the stub-out, threads outward. Apply pipe-welding solvent to the

plastic ferrule and slip it onto the stub. The ferrule's tapered side should face the wall. When the solvent has fully hardened, the angle-stop can be installed by tightening its flange nut over the body of the angle-stop. If your supplier doesn't have this specialized fitting, you can install a plastic male threaded adapter and use a regular angle-stop with female pipe threads.

Fixture straight-stop — Installation of a straight-stop is the same as for an angle-stop, but a straight-stop fixture fitting is designed for use with floor stub outs.

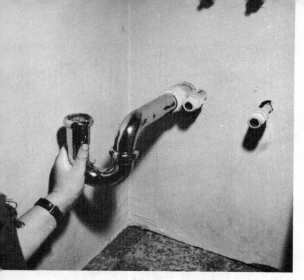

Chromed fixture P-trap slips into branch drain stub-out and swivels to meet the fixture drain. When the slip-nuts are tightened the pipe becomes watertight.

Get speedees that are long enough to reach from the floor to the faucet. Three-footers should do it.

Compression coupling — If you are going economy class, you can avoid paying for the stop valve feature by threading a fixture compression coupling onto the stub-out. There are ⅜×⅜-inch and ½ ×⅜-inch sizes. Both straight-through and right-angle types are made in male and female threads. Use the right-angle ones for wall stub-outs, the straight-through for floor stubs. Your speedee will fit into the ⅜-inch compression fitting end.

Hooking up the fixture end of the speedee is done with a coupling nut that draws the flanged or bayonet end of the speedee into the tailpiece of the faucet supply fitting. If the faucet comes with short pipe-threaded tailpieces, throw them away.

If the fixture faucet has soft copper water tubes, you can run them directly into the compression coupling on a wall angle-fitting. If they're too short to reach, you'll have to put a compression or flare coupling between them and one end of a speedee. A plain end can be made on a flanged or bayonet-ended speedee sim-

ply by cutting it off. The end can then be flared easily, if desired.

CONNECTING THE WASTE

Every fixture except a toilet must be fitted with an external trap. Traps on lavatory, sink and laundry tubs, where underneath looks is important, should be made of chromed brass. Use 1¼-inch for lavatories, 1½-inch for other fixtures. A *P*-shaped trap is used for wall stub-outs. An S-shaped trap is used for floor stubs. The trap is made in two pieces so it can be swiveled to fit various fixture-waste pipe alignments. The waste pipe stub should reach ½ inch beyond the finished wall or ½ inch inside a cabinet-type fixture. Install the right length nipple in a steel supply system or cut off a copper or plastic nipple to the right length. End up with a 1½-inch-diameter male-threaded stub-out projecting ½ inch out of the wall or cabinet back.

Fixture drains come down from the drain fitting at the bottom of the bowl with 4-inch tailpipes. These are screwed in. You can get 8-inchers too, if you need to reach a low trap. The drain fitting itself is held with a rubber washer, metal washer and large fixture nut. Unless two large rubber washers are included, put a ring of plumber's putty around the top of the drain fitting before you install it. When the drain fitting is tightened, excess putty will squeeze out. Put pipe joint compound on the threads of the tailpipe before you screw it in.

To install a chrome-plated fixture trap, place the escutcheon plate over the stub-out. Then slip the trap's end leg into the stub-out. Fasten it loosely with a slip-nut and rubber washer. These come with the trap. Slip the other trap leg over the tailpipe and secure it loosely with a slip-nut and rubber washer. Make sure that both trap legs are square with their pipes at both ends, then tighten the slip-nuts with a monkey wrench or fixture wrench. Tighten the swivel fitting between legs of the trap, too. If you're going first class, buy traps with drains for easy clean-out.

A copper waste system can be fitted with what are called *marvel couplings*. These take the place of stub-out pipes letting you slip a 1¼- or 1½-inch fixture trap directly into them. They have built-in slip-nuts and brass ferrules to grip the trap. Order by size.

Traps that don't show, such as for kitchen cabinet sinks, can be made of less costly materials. A sink trap for hidden use often has a malleable elbow fitting at the end instead of a chromed brass pipe. The elbow screws onto the stub-out.

Laundry tubs are generally fitted with 1½-inch galvanized *P*-traps. The drain fitting inside each tub is screwed to the top of the *P*-trap. With copper or plastic, a male threaded adapter is needed at this point. A downward-pitched horizontal run is made from the trap to the branch waste.

All fixtures should be handled carefully to avoid chipping or scratching. Never slide them across concrete or tile floors. Use pads to rest them on and wooden blocks to brace them into position. Don't stand in a fixture with your shoes on. Stand in stocking feet or on a newspaper cushion.

The piping parts that are exposed beneath the fixture are considered a part of it, though they may not come with the fixture. They're usually chrome plated. Don't forget to order them when you order the fixture: things like traps, supply pipes. Order faucets too. If they're one-piece units, they must have centers that match those of the fixture. Standard on lavatories are 4-inch centers; 6- and 8-inch centers are common on sinks.

Enough general information about fixture supply and waste connections. Here are some specifics to guide you in your installation.

LAVATORIES

Lavatories are designed to be wall-

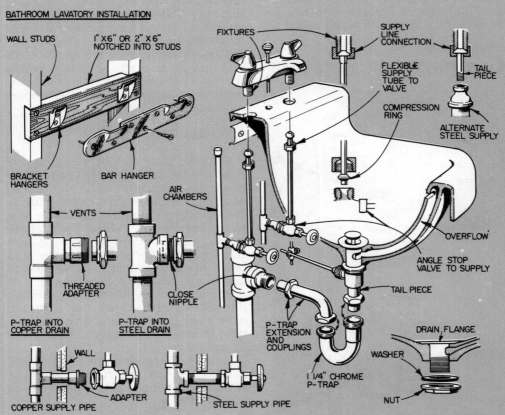

BATHROOM LAVATORY INSTALLATION

WALL STUDS

1" X 6" OR 2" X 6" NOTCHED INTO STUDS

FIXTURES

SUPPLY LINE CONNECTION

FLEXIBLE SUPPLY TUBE TO VALVE

TAIL PIECE

COMPRESSION RING

ALTERNATE STEEL SUPPLY

BRACKET HANGERS

BAR HANGER

AIR CHAMBERS

VENTS

THREADED ADAPTER

CLOSE NIPPLE

OVERFLOW

ANGLE STOP VALVE TO SUPPLY

TAIL PIECE

P-TRAP INTO COPPER DRAIN

P-TRAP INTO STEEL DRAIN

P-TRAP EXTENSION AND COUPLINGS

1 1/4" CHROME P-TRAP

DRAIN FLANGE

WASHER

NUT

WALL

ADAPTER

COPPER SUPPLY PIPE

STEEL SUPPLY PIPE

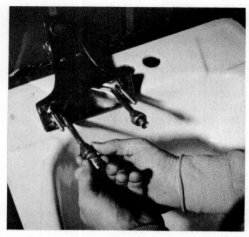

To install a lavatory faucet extend the soft copper tailpipes with flare fittings so that they reach the transition fittings.

Put a ring of plumber's putty on fixture bowl and install faucet unit. Clean up excess putty after tightening the faucet.

hung, pedestal- or leg-supported or be built into a cabinet. The top of a lavatory is about 31 inches above finished floor level. Wall-hung lavatories are fastened onto metal brackets attached to the wall. Provide a 1×6 board nailed to the studs behind the lavatory and flush with the wall surface. Screw the bracket to the board. Install the faucet unit before putting the fixture up. It's easier that way. Put a ring of plumber's putty around the faucet base before you insert it through the fixture.

If there are legs for the lavatory, adjust them to level the bowl top. Follow the manufacturer's instructions in hooking up and adjusting the fixture stopper mechanism.

KITCHEN SINK

A kitchen sink may be wall-hung, but most are the cabinet-type. The bowl, or bowls, must be installed in the cabinet according to the instructions with them. Before you position the cabinet be sure to cut holes in the cabinet back for the supply pipes and fixture drain. Attach the faucets before installing the sink bowl. Make the supply and waste connections the same as for a lavatory.

When a garbage disposal is used on one side of a two-bowl sink, it should have a separate trap and waste-connection. Otherwise it's likely to back water up into the opposite bowl.

Instead of speedees for piping water to a kitchen sink, you can use lengths of ½-inch soft copper pipe with threaded adapers.

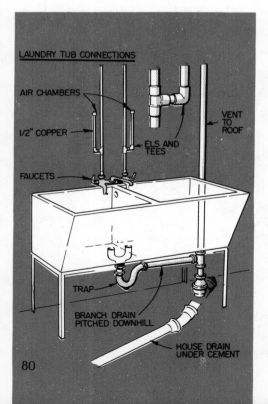

LAUNDRY TUB CONNECTIONS

AIR CHAMBERS

VENT TO ROOF

1/2" COPPER

ELS AND TEES

FAUCETS

TRAP

BRANCH DRAIN PITCHED DOWNHILL

HOUSE DRAIN UNDER CEMENT

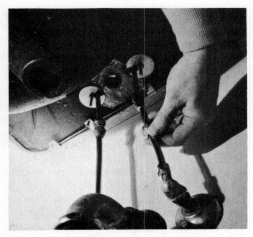

Flared speedees carry water from transition fittings on stub-outs to flare couplings on the ends of the faucet tailpieces.

Spread putty around the drain and install the drain fitting and collar. Some drains have a rubber washer. Don't putty them.

A laundry tub is a switcheroo on a sink. The supply pipes for a basement laundry tub are usually brought down from above to meet the faucet connections. Air chambers can be made by installing tees just above the faucet connections. Go out with short nipples, then up with 90-degree elbows. Each air chamber fits into an elbow and is made of a 12-inch-long capped nipple.

BATHTUB

A bathtub is different, too. This is the only fixture that should be installed before the walls and floors are finished. Boards, 1×4-inchers, are nailed around the tub on three sides just high enough that the tub flanges can rest on them. No boards are needed for cast iron tubs. The tub covers its framed opening from end to end. Follow the rough-in measurements given by the tub manufacturer.

Under some codes you must have access to a tub's waste and supply piping. A first floor tub can have access through a 12-inch-long, 6-inch-wide opening cut through the floor at the head of the tub. The opening should be centered on the tub and extended 3 inches into the area below the wall in front of the tub. Access for a second floor tub can be provided by a removable panel on the wall opposite the tub's head. With access provided, the

tub can be placed in its opening and the walls and floor finished.

More modern codes permit tub hookup with access only through a large circular opening behind the faucet plate. This would permit replacement of the

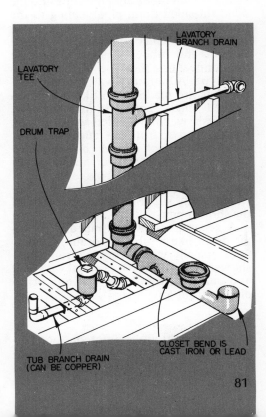

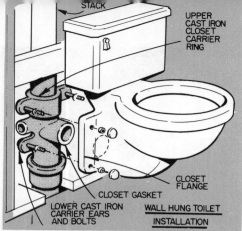

Labels in image:
STACK
UPPER CAST IRON CLOSET CARRIER RING
CLOSET FLANGE
CLOSET GASKET
LOWER CAST IRON CARRIER EARS AND BOLTS
WALL HUNG TOILET INSTALLATION

Bathtub branch waste run.

faucet. A *running trap,* not a drum- or P-trap, is used. It starts horizontally, dips down, comes back up and ends horizontally again. No access is needed. If it ever clogs, it can be cleaned with a flexible plumber's auger run in through the tub overflow pipe.

Install the tub fittings and overflow fittings to the tub. These are threaded to screw into the special brass drainage fitting that comes with the tub fittings. The 1½-inch tub drain pipe slips into the male threaded drain pipe leading to the trap. For an accessible hook-up you can secure the connection with a 1½-inch slip-nut and rubber washer. Otherwise the connections must be of the permanent type. A bathtub's water supply connections are made much like those of a kitchen sink. They're ½-inch, usually.

If a shower is included with the tub, this is piped up from the faucet's shower outlet with ½-inch pipe. A chrome-plated shower pipe with shower head reaches out from a 90-degree elbow at the top of the shower pipe. A 1×4 board set between studs supports the upper end of the shower pipe.

A separate shower is piped with water much like a tub. The drain is underfloor, usually with a running trap. Most shower floor fittings are calked with lead and oakum around the 1½-inch drain pipe to form a tight seal.

TOILET

A toilet hook-up is different, too. Toilets have built-in traps, no separate trap being needed. They set right down on top of a floor flange connected to the toilet drain. It's important that this floor flange be properly positioned for the toilet you're using. For most toilets its center should be 12 inches out from the *finished wall* surface. This is called the toilet's *rough-in* dimension. If you measure only that far from the stud wall for a rough-in, you're in trouble. Some toilets call for a 12½-inch rough-in. Others need a 13-inch rough-in. Too much is better than too little. A toilet can be put in with additional space behind it; however, if there isn't enough space, something's got to give. Check the rough-in for the toilet you're going to use before installing the floor flange and closet bend.

A wall-mounted toilet may be preferred for its modern appearance. Wall-mounters are easy to clean under too. Some have their tanks hidden in the wall. Most require a 2×6 stud wall. Some also employ a heavy metal carrier that must be securely fastened to the framing or to the main stack at the right height. An access panel must be provided for servicing an in-the-wall tank.

Whether the toilet's branch drain is plastic, copper, cast iron or steel, its floor flange must be flush with the floor, or not more than ½ inch below it. The bottom of the toilet bowl has a ridge around its drain opening. The ridge rests on the floor flange with a wax or rubber gasket, or a ring of plumber's putty to seal the joint. Bolts in slots around the floor flange project up through holes in the bottom of the base of the toilet bowl. These are tightened to secure the connection. In some cases, wood screws are turned down into a wood subfloor instead. Plastic or china covers are placed over the bolts.

A wax or rubber toilet bowl gasket is preferable to plumber's putty for setting the bowl. Install the gasket over the rim with the toilet bowl resting upside down on pads. Then invert the bowl and gently lower it straight down onto the floor flange. Press down on the bowl with your full weight and twist it to settle it over the gasketed connection. The bowl should end up exactly level when it's all the way down on the floor. Check it with a level. Install wedges if needed, but be sure that wedging doesn't lift the bowl enough to leave a gap in the gasket connection. Get the bowl square with the wall behind and draw up the bolts. Don't force them. You can crack the bowl. You may want to put plumber's putty around the base of the bowl for a neater installation.

Unless yours is a one-piece toilet, attach the flush tank to the back of the bowl according to instructions with the unit. Don't force these bolts either. Connect the toilet water supply line just as for a lavatory, using an angle-stop or straight-stop and a toilet-type speedee. When you screw the supply line fitting onto the end of the tank ball cock, hold the ball cock with pliers to keep it from turning.

DISHWASHER

An automatic dishwasher is a fixture of sorts. Follow the piping instructions with the one you're installing. A dishwasher is usually fitted with an air gap drain that prevents kitchen sink drainings from backing up into the dishwasher tub. An air gap device also avoids violating provisions in some codes that prevent pumping of water into a drain. The air gap unit usually comes with the dishwasher. Like other fixtures, a dishwasher should have a trap all its own, placed after the air gap. It needs its own branch drain, too. Don't pipe a dishwasher into a kitchen sink drain unless it's a portable dishwasher. That's a different animal.

AUTOMATIC WASHER

An automatic clothes washer often needs supply and waste piping of its own not connected with an adjoining laundry tub. Run hot and cold supply lines to it. Create extra-large air chambers by teeing off and then up with ¾×18-inch pipes. The extra large air chambers take up sudden shocks of quick-closing solenoid washer valves. Provide stop valves on the supply lines at the ends just before the final hose bib fittings. The washer's hoses attach to the bib's fittings. A

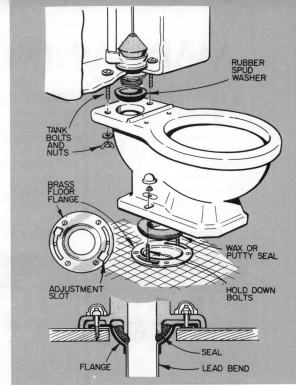

Mounting a toilet and close-coupled tank.

Newest type of toilet bowl gasket is wax with plastic flange for complete sealing.

washer's supply pipes may be fastened to a basement wall or aimed out the side of an enclosed laundry tub cabinet.

A waste for the washer should contain a 1½-inch P-trap on the end of a pipe leading to the main stack or other drain. Screw a 12- to 18-inch nipple into the top of the P-trap as a standpipe. The washer drain hose may be slipped down into this.

Asmall improvement in the efficiency of a heating plant can bring big savings in fuel bills over a year's time. This is true whether the heating plant is steam, water or air; gas, oil or electric. For this reason—for safety too—most heating experts recommend an annual inspection of the whole heating system. A good time to do it is well before the heating season is on. Then if any repairs or adjustments are needed, you have time.

Your central heating system consists of five separate systems: (1) Fuel burner. (2) Furnace or boiler. (3) Heat distribution system (ducts, pipes). (4) Room heating units (radiators, registers, convectors, etc.). (5) Controls (thermostat, dampers, pump, blower, etc.). All must

Replace dirty air filters at least twice a year to ensure proper furnace operation.
Lennox Industries

A periodic check-up will keep your heating system operating at its best

be in good order if you're to get efficient heat.

An obvious starting point is the heating plant. While you are working around the heating plant, clean it inside with a vacuum cleaner. Soot and scale are insulators that can keep combustion heat from getting through to the house. This wastes fuel up the chimney. Remove the access hatch. Use the brush attachment of your vacuum to clean soot deposits from the inside of the combustion chamber. Boiler scale can be attacked with a stiff-bristle wire brush. The longer the handle the better.

CHECKLIST

Lubrication—Motor, fan or water pump bearings may need lubrication. Those that aren't self-lubing should have their cups or filling tubes oiled at least twice a year with the lubricant recommended by the manufacturer. Usually this is a good grade of 30-weight motor oil. Make sure the furnace switch is turned off when you oil.

Belt adjustment—When you lubricate is the best time to check and adjust tension of the fan belt on a forced-air furnace. Turn off the switch and tighten a loose belt by turning the cradle bolt beneath the blower motor. The belt should have 3/4 to 1 inch of play midway between pulleys. A belt that's too loose may slip and wear out quickly. At the same time it may not drive the blower as fast as it should. A belt that's too tight wears out belt, motor and fan bearings faster than a properly adjusted belt does.

A belt can cause creaking or squealing noises in the heating system when it reaches the worn out stage. Any belt that is noisy or is cracked or frayed should be replaced.

While you're in this section, vacuum the blower's squirrel cage to remove accumulated dust.

Clean the filter—Nearly all forced-air heating systems use filters to trap dust in the moving air. Most of these are of the

Unless self-lubrication, fill oil cups on blower and oil pumps with oil twice a year.

Give blower belt ¾-inch of play between pulleys. A too-tight belt wears out fast.

Lennox Industries

Make sure collars on blower assembly are tight. Turn set screw with allen wrench.

Vacuum fins of blower cage to restore efficiency and eliminate circulating dust.

Leave burner service to a technician. Do-it-yourself here can be dangerous to you.
Lennox Industries

replaceable glassfiber type in which the fibers are coated with an oil that grabs and holds tiny bits of dust. However, after the fibers have trapped all the dust they can hold, the rest of the dust can go right on by. An air filter in this condition can only trap large lint particles. What's more, air movement is restricted, cutting down on the amount of air circulation your heating plant can provide.

Unless you have a means of re-oiling a glassfiber filter, don't try to clean one. Consider it used up and get a new filter element.

You should change filters every month during the heating season in cold climates. Install them with the arrows on the frames pointing in the direction of air flow. Neglected filters cause more complaints of poor heating than any other single factor.

Check breathing—Your furnace or boiler must have air, just like you need air to breathe. The biggest danger is that in a tightly sealed home your burner may become oxygen-starved. This can result in lethal gases being released into the house. Whether the heating plant is located in a closet, basement, attic or crawlspace, provide some ventilation. Have a screened vent, a door with grille in it or, in the basement, leave a window open about an inch. The storm sash could be left off a loose basement window to provide combustion air to a basement burner.

If you smell burning fuel and can't get rid of it by providing ventilation, call your serviceman fast.

Air that gets pushed out by a kitchen or bath vent fan or outside-vented clothes drier must be replaced. The same is true of fireplace draft air. Keep adequate breathing air for your heating plant by having a window open several inches while these are operating.

Pilot light—In spite of what you've heard about turning off a gas heating plant's pilot light during the off-season, you won't save money by it. Heat from the tiny pilot keeps your furnace dry inside, helping to prevent corrosion.

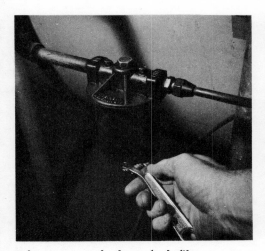

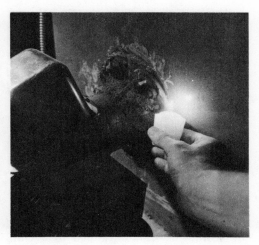

Take apart and clean fuel filter once a year for free flow of fuel during winter.

Find unwanted leaks in firebox of burner with a candle. The flame follows drafts.

OIL BURNERS

Most old oil burners have fireboxes too large for the capacity of their burners. You can check yours by measuring the area of the firebox in square inches. If the area is more than 75 square inches for each gallon-per-hour capacity of the burner nozzle, you should have a smaller box. Your present firing box is easily made smaller by installing a liner kit. Johns-Manville makes one called *Cerra Form* out of ceramic-fiber panels. These assemble into a smaller firebox inside your present one.

As part of your annual inspection of an oil burner, have a combustion efficiency check made by your serviceman. The efficiency should read 75 percent or more. Have him check the baffles in the firebox. Are they in place and are there enough of them? Air leaks into the firebox should be sealed off with calking.

You can inspect the burner nozzle to see that it's the right size for the diameter of your firebox. Check it against these ideal chamber diameter-nozzle gallons per hour combinations:

    10-inch— .50 to .85 gph.
    11-inch—1.00 to 1.25 gph.
    12-inch—1.35 to 1.50 gph.
    13-inch—1.65 to 1.75 gph.

Flush out the fuel oil strainer to get rid of dirt collected by it. If there isn't one, you'd do well to install one between the fuel tank and burner.

If your oil burner has no draft stabilizer, you are likely losing too much heat up the chimney. Install one.

With any burner, a good chimney should be clean and free of obstructions. Knock soot accumulations from the chimney's inside walls by lowering a brick tied to a string. Afterward debris should be removed from the clean-out opening at the bottom of the chimney. Lacking a clean-out opening, you'll have to pull off the vent pipe from your heating unit and remove the soot there.

Some things should be left to the pro's. Don't fool with the pilot flame and safety controls on a gas burner. Never mess with the ignition parts on an oil burner. Leave the oil pump alone too. Your serviceman should do the checking of the safety valve on a steam boiler and the pressure relief valve on a hydronic system. Your family's safety is at stake.

HEAT DISTRIBUTION SYSTEM

What you can do to your heating distribution system for increased efficiency depends on what type of system it is—

Calk crack around oil burner to seal off efficiency-spoiling air leaks in the firebox.

Knock off soot collecting on chimney wall by scuffing it with brick tied to a cord.

With inlet valve on, to admit fresh water to system, flush dirty water from boiler.

air, water, steam or electric. Radiant electric heating components require little maintenance. Electric furnace types, in which air heated by electricity circulates through ducts, need the same attention any forced-air distribution system needs.

Ducts — Ducts and pipes running through unheated spaces should be insulated. Heat loss into such areas, ups your fuel bill. In basements where the heat is utilized, you needn't insulate. Return runs may not need insulation in milder climates.

Ducts should be calked at joints where escaping heat would be wasted. Use calk that sticks tightly to metal and can take heat. Both silicone calks and the new acrylic calks work well.

Drain expansion tank — Every modern hydronic heating system makes use of an expansion tank partly filled with water. As the water in a closed system is heated, it expands against a cushion of air. Some of the air becomes dissolved in water, and the tank gets waterlogged. Its air cushion is depleted. For that reason an expansion tank should be drained every year to restore the full air cushion.

Flush your system at the same time. Open the drain valve at the bottom of the boiler and flush until clear water comes

Drain expansion tank to replace its air-charge during heating system inspection.

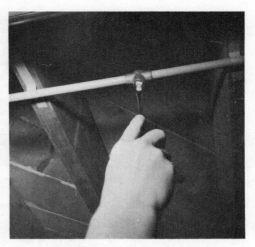

Vacuum air registers to remove accumulated dirt and dust, using special attachment.

Flow valves balance hydronic heating systems. Screw slot shows angle of the valve.

out. Then refill to the proper point on the gauge.

Vent radiators — Air collects at the tops of radiators in a hydronic system and should be bled off every year. A radiator that's half filled with air is only half a radiator. Open the vent on each unit until all the air escapes. Then close it firmly.

Radiators, convectors, registers — Heating is efficient only when air can freely circulate around and through a radiator or convector. Radiator covers and marble or wood shelves on top of radiators, block air circulation. If you must use a radiator cover, be sure the design is correct. Ample space should be provided at the bottom for cool air to enter and at the top for hot air to get out.

Also be sure that draperies, rugs and furniture don't restrict heat from around registers and radiators. Dust or dirt on registers and convectors prevents free movement of heated air. Keep them clean. Most vacuum cleaners have an attachment for this purpose.

One trick to help a convector pour out heat is to paint the back of its cover flat black. This will make it transmit heat like a radiator. You can do this easily while you have the cover off. Vacuum the fins too.

COMFORT ADJUSTMENTS

Whether you have forced-air or hydronic heating, your heating system should be adjusted to provide constant circulation of the heating medium. This is done for comfort. The blower or pump is adjusted to run all the while the outdoor temperature is below 45 degrees. This gives you tablespoon quantities of heat continuously, not bucketsful every hour, as other adjustments give.

Constant circulation is accomplished by adjusting the blower or pump limit control settings and sometimes the flow of air/water. Your heating serviceman can make the adjustments for you.

BALANCED HEATING

Chances are three out of four that your heating system is out of balance. Many are. Unbalance leaves some rooms too warm, others too cool. This is true even of new houses. Whether your system is hydronic or forced-air, balancing is done in much the same manner. Balancing the system yourself in not complicated, in fact you can probably do as good a job as a professional. You can spend the necessary time, without call-backs or wasting of time waiting for the system to normalize after each adjustment.

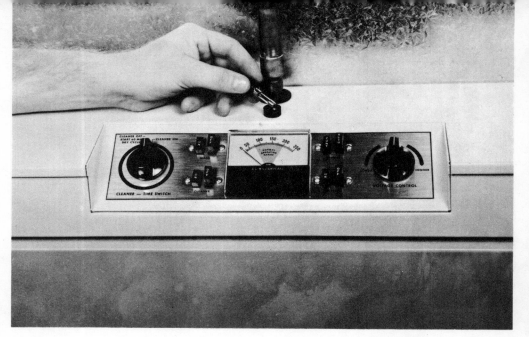

All you can do to service an electronic air cleaner is to see if the fuse has blown.

Make a balance adjustment during cold weather when the sun is not shining so that the heating system has a good load on it. Balance is achieved by restricting the flow of heat to rooms that need more. Of course, the room where the thermostat is located should stay at proper temperature all the time.

Before making a heat balance adjustment, storm windows should be installed, filters should be cleaned, expansion tank drained, air bled from radiators and the blower or water pump operating properly. All windows and doors that would normally be closed when maximum comfort is required should be closed.

Begin balancing by opening all the restrictions in ducts or pipes. In ducts these are in form of dampers. There is normally a damper in each individual duct run. Sometimes control of heat is at the registers. Check these too. Usually each control has a screw slot or other indicator showing the damper- or valve-position. Full flow is when the screw slot is parallel with the flow. In hydronic systems the restrictors are valves located in supply pipe runs where they tee off to a radiator.

Carry an accurate thermometer around to each room. Check the temperature about four feet from the floor. Record the readings. Then you'll see which rooms are too warm and which are too cool. They should all be within 2 degrees of the thermostat setting.

Close the dampers or valves slightly in runs leading to rooms that are too warm. Changing the flow to one room will usually affect the flow to all rooms, so don't over-adjust on the first try. Reducing the flow of heat to warm rooms should provide adequate heat to cool rooms. If they're still too cool, further reduce heat flow to the other rooms or restrict the input of heat to the room containing the thermostat.

Be sure to allow ample time after a change in adjustment for the rooms to reflect it fully in a temperature change. Forced-air systems react faster to a change than hydronic systems do. Radiant systems are slowest to react. In a radiant system make your next check the following day.

Mark the damper positions for future reference.

Your thermostat may need adjusting of its heater scale to give proper burner cycling. Have a serviceman do it.

iber glass on top of boiler or furnace
lenum chamber saves heat by insulation.

You can make obvious repair to air conditioning system, like tightening fittings.

TYPES OF TOILETS

Understand the difference in this fixture before you buy for your home

Is there a difference in toilets? You bet there is. A BIG difference. And you get about what you pay for. Better toilets are cleaner, more sanitary, more comfortable, more efficient, quieter and more effective in the way they flush.

There are three toilet types (in ascending order of cost): siphon washdown, reverse trap, siphon jet. When the handle of a toilet is tripped, four to 10 gallons of water cascade down from the toilet tank. The water gushes out from small holes around the toilet bowl's perimeter. As water fills the bowl, it builds up a head of pressure that starts water moving over a lip at the rear of the trap. The trap has a smaller outlet than inlet. What's more, there are turns and offsets in the outlet leg of the trap so that water will flow in faster than it flows out. This fills the trap with water.

Once the passages of the trap are filled, a siphon action starts that produces a partial vacuum. Atmospheric pressure forces water out of the bowl. The flushing action continues until the water level falls below the lip at the front of the bowl.

Then air is sucked into the trap, quickly breaking up the siphon.

MAKE-UP WATER

Since a flush leaves the toilet bowl virtually empty and the toilet a vulnerable spot for sewer gas to enter the house, the bowl must be filled again. To do this a refill tube in the toilet tank spews water into the overflow pipe all the while the tank is filling. The trap is resealed.

A siphon washdown toilet, *washdown* for short, is the cheapest and least desirable toilet type. Depending on what brand you buy, a siphon jet toilet costs from $5 more to twice as much as a washdown. Unless a low first cost is your only goal, forget a washdown toilet. You won't like it. Its small trap passages coupled with a so-so flushing action make it slow to flush and quick to stop up. It uses more water, too. This is big drawback in a water-short area or where a home disposal system is already in enough trouble. Because all the flush water must pass through the bowl, a

A modern wall-hung toilet facilitates cleaning the floor, permits wall-to-wall carpeting of the bathroom, gives a more spacious feeling and enhances the bathroom decor.

washdown is noisier and splashier than the other types. Its smaller water surface leaves a larger area of the bowl dry above the water line. The bowl is thus harder to keep clean.

REVERSE TRAP

Though the reverse trap has an action like a washdown, its rear trap permits a larger water area and deeper trap seal. A reverse trap toilet isn't the manufacturer's cheapest model, and so is usually available in colors.

A siphon jet is built much like a reverse trap toilet, but with a built-in jet. The jet is a 3/8-inch stream of water that shoots directly into the trap to whip up action in a hurry. If you can afford a siphon jet, get it.

A siphon whirlpool, which spins the water like a tornado, is a more recent development. It isn't available from every manufacturer. Those who make it claim a positive, thorough flush with less rushing water noise. As with all top-of-the-line products, siphon jet and whirlpool toilets are available in the latest and best looking designs.

Practically all modern toilets are built with close-coupled tanks and bowls. The most modern ones have one-piece tank-bowls. These cannot overflow. Some two-part toilets have overflow-prevention too. By all means get a toilet with a more sanitary elongated bowl. They're available in all types but washdown.

Bowl height is coming down. Originally it was 16 inches. Now most toilets are 15 inches high. Some only 14 inches. Some toilets feature self-ventilating bowls.

Toilet tanks have been improved too. The rubber flapper valve has largely replaced the old problem of toilets, the flush ball. Tilting buckets and other types are popular too. As a result, the newer toilets are much less prone to trouble than the older ones were. That insidious cross-connection, the submerged toilet tank inlet, has been replaced by anti-siphon valves.

Know the functional features you want in a toilet. Then shop for manufacturer, style and price.

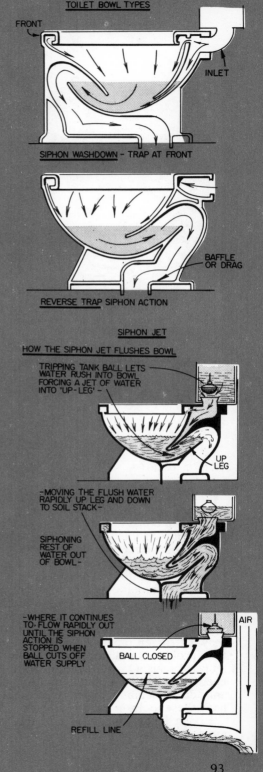

TOILET BOWL TYPES

FRONT

INLET

SIPHON WASHDOWN - TRAP AT FRONT

BAFFLE OR DRAG

REVERSE TRAP SIPHON ACTION

SIPHON JET

HOW THE SIPHON JET FLUSHES BOWL

TRIPPING TANK BALL LETS WATER RUSH INTO BOWL FORCING A JET OF WATER INTO 'UP-LEG' -

UP LEG

-MOVING THE FLUSH WATER RAPIDLY UP LEG AND DOWN TO SOIL STACK-

SIPHONING REST OF WATER OUT OF BOWL-

-WHERE IT CONTINUES TO FLOW RAPIDLY OUT UNTIL THE SIPHON ACTION IS STOPPED WHEN BALL CUTS OFF WATER SUPPLY

AIR

BALL CLOSED

REFILL LINE

Repairing a leaky faucet is something that every home plumber should be able to do.

MAINTAINING
HOME PLUMBING

Make repairs on faucets and valves; stop leaks in toilet bowl seals

The cost of calling a plumber keeps going up. His skill and scarcity demands top dollar in today's market. Consider yourself lucky if you can even get one to come out on short notice. To save some bucks and inconvenience, it may be worthwhile to tackle the simpler of your own plumbing repairs. Jobs that you do with a few inexpensive tools include repairing leaky faucets, clearing clogged drains, fixing leaks in pipes and tanks, thawing frozen pipes, fixing balky toilet tank problems and stopping leaks in toilet bowl seals. Only a few basic tools are needed, including a plumber's force cup and drain-cleaning auger. You should have these tools on hand.

Fixing leaky faucets is probably the most basic home plumbing repair. It's not as easy as it used to be when faucets were built pretty much alike. Several improved designs have gained wide acceptance. However, if you know what you're doing, you can fix them, too. In 10 minutes you can repair a leaky faucet of the ordinary type.

The cure is worth the effort. A faucet that leaks just 60 drops a minute wastes 2300 gallons of water in a year.

Actually, water faucets and globe valves do the same job. The difference, mainly, is that faucets are valves placed at discharge points over sinks, tubs and lavatories. Valves are used to close off portions of the plumbing system. The repairs to a faucet apply also to a globe valve. Mixing faucets found on sinks, laundry tubs and bathtubs are actually two separate faucet units with a common spout. Each faucet is repaired individually.

FIXING A GLOBE FAUCET

When a closed globe faucet drips and "sings" or "flutters" when opened, the trouble is usually a worn washer. Don't force it closed to stop the drip. Replace the faulty washer instead. The faucet washer is located at the lower end of a spindle, the part that turns with the handle. To replace the washer, turn off the water supply to that faucet. Remove the packing nut, then the spindle. Wrap adhesive or other sticky tape around the packing nut to protect the chrome finish. Use a monkey wrench with smooth jaws, not a pipe wrench with toothed jaws. Turn the spindle in the "on" direction to remove it.

Take out the small screw that holds the worn out washer to the spindle. Scrape the old washer parts from the spindle cup and insert a new washer the same size as the old one. Buy washers that will serve both hot and cold water taps.

If you don't have a washer the same size, you can make one by filing down a larger one. Chucking it up on a bolt in an electric drill is an easy way to let power do the work. Never use a faucet washer that's too small.

Look at the seat on the faucet body. If it's nicked or rough, reface it with a seat-dressing tool. Then reassemble the faucet. Don't forget to get the "hot" handle on the "hot" faucet. Don't laugh. It's easy to go wrong.

Sometimes a faucet leaks at the packing. Try tightening the packing nut. If that doesn't work, you'll have to replace the packing. To do it, just remove the handle, packing nut and old packing. Install the new packing or packing washer and reassemble the parts. Other globe faucet or valve parts may be replaced as needed.

Emergency packing can be made by wrapping cotton string tightly around the faucet stem just inside the packing nut. Wrap so that the string follows the same path as the handle's threads.

The final step is to remove and realign the handle to match the one opposite. Most handles are splined onto the shaft and held with a screw.

Cross-section of a typical globe faucet.

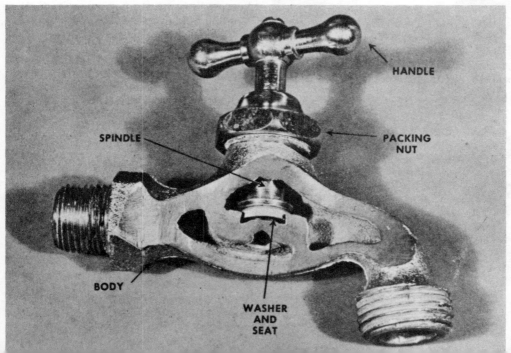

To fix leaky washer, unscrew packing nut on globe faucet. Use tape to protect nut.

Lift out spindle, turning it slightly as you do. The water should be turned off, of course.

A screw holds faucet washer to bottom of spindle. Remove to install a new washer.

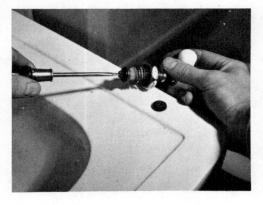

FIXING OTHER FAUCETS

Some of the newer faucets don't have washers. Instead they have diaphragms that close off the water supply. They're just as easy to fix. Remove the handle. The screw that holds it usually is under a snap-plate covering the center of the handle. Pry the plate off with a pocket-knife. With the handle off, screw the hex nut around the spindle off, and pull out the spindle assembly. Replace the neo-prene diaphragm with a new one and re-assemble the faucet. There is no packing on this type of faucet. If it leaks around the shaft, a new diaphragm is the cure.

Other kinds of faucets are designed to eliminate drip and prolong life. Consult the service literature provided for these when repairing them.

Single-lever faucets operate differently from other types. They use spring pressure along with water pressure for shut-off. These are particularly affected by pipe sediment, and so are protected with tiny metal screens to filter out debris. When the screens clog up, they cut down the flow. You have to get at them for cleaning. Other than this, a single-lever doesn't need much attention.

Every make of single-lever faucet is a little different. The best bet is to come by a copy of the manufacturer's literature telling how to service it. Call a plumbing dealer who handles that make and ask him for literature. Or write the factory and get it, along with the name of the dealer where you can buy parts. Send along a snapshot or description of the faucet, so the maker knows which of his models you're talking about.

CLOGGED DRAINS

Drains get clogged by dropping things into them, or by build-ups of grease, dirt and other matter. First try plunging out a stopped-up drain with a plumber's force cup (*plumber's friend*). Fill the bowl about half full of water and roll the plumber's friend into the water so that no air is trapped under it. If the fixture has an overflow opening, plug the opening with a damp rag or hold one hand over it as you plunge. On a double-bowl sink, have someone hold a hand tightly over the opposite drain. Remove the stopper from the drain you are plunging. Give it 20 to 30 good shots, not just a few. Getting the right rhythm can send a powerful shot down the pipes each time you plunge. Every so often, yank the force cup off the drain to give additional plunging action and test whether the drain has been cleared. Plunging action may be increased by coating the bottom lip of the force cup with vaseline. It makes for a tighter seal.

Once the drain has been opened, if only slightly, you can pour liquid drain cleaner down to clear it the rest of the way. Follow directions on the label.

You can make your own drain cleaner by mixing lye with a small amount of aluminum filings. Handle it with care. Pour it into the drain and add cold water. A violent heat- and gas-forming reaction should loosen remaining grease and soap deposits. Then they can be flushed away. Keep your hands out of drain cleaner. It's mean stuff. If any gets into your eyes, flush with cold water and call a doctor. Don't use chemical cleaners in pipes that are completely stopped up. The chemical troops must engage the stoppage in a frontal attack to be effective.

Stubborn deposits that won't come out with a few minutes of vigorous plunging will have to be removed with a clean-out auger or sewer tape. Long flexible steel cables or ribbons, these are commonly called *snakes*. They are run down drainpipes as they're rotated slowly to break up obstructions or hook onto objects and pull them out. They're made in various lengths and sizes and are available at hardware stores and plumbing supply houses. A 10-foot snake is a handy size. It can reach through most drain pipes clear to the main stack. Rarely is the main stack clogged. If the snake hangs up, don't force it. Back it out a few inches and go in again. Try slow. Try fast. Try not turning. Try fast turning. Chances are, you can get it through, if you keep trying.

Water pressure from a hose is a slick way to blow out some stoppages. Insert the hose end well into the pipe. Wrap rags around the hose where it enters the pipe to minimize backflow of water. In case of a stopped-up fixture drain, the overflow opening has to be covered tightly too.

If the stopper in a sink or lavatory leaks, it's probably dirty. Take it out and clean it. Most simply lift out. Some call for cleaning from below with the trap and tailpipe off.

Plug overflow opening with hand when using a plumber's friend on stopped-up lavatory.

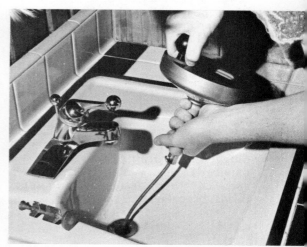

Turn plumber's "snake" a few inches at a time while feeding it into clogged drain.

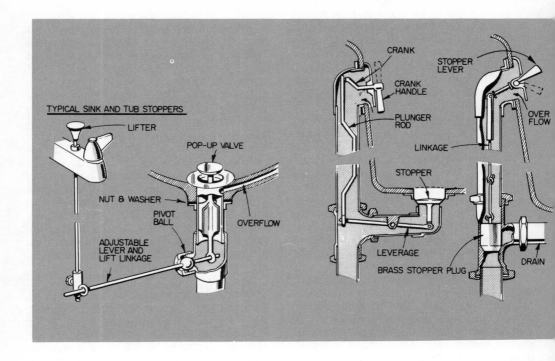

TYPICAL SINK AND TUB STOPPERS

LIFTER

POP-UP VALVE

NUT & WASHER

PIVOT BALL

OVERFLOW

ADJUSTABLE LEVER AND LIFT LINKAGE

CRANK

CRANK HANDLE

STOPPER LEVER

PLUNGER ROD

OVER FLOW

LINKAGE

STOPPER

LEVERAGE

BRASS STOPPER PLUG

DRAIN

CLOGGED SEWER LINE

A clogged sewer line is cleaned out with a sewer tape, a flat metal band that comes coiled in a carrier. You can rent sewer tapes in lengths of 25 and 50 feet. Work the tape through a clean out opening after draining backed-up water into pails and disposing of it. Uncoil the tape from its rack as you plunge it into the clogged line. When you clear out the stoppage, work the tape back and forth over the area a few times. Then reel it in. Clean the tape and dry it before recoiling and returning.

A clogged toilet usually can be cleared with a plumber's friend used as for a clogged sink drain. Next in line is a plumber's snake. A short model is made specifically for clearing clogged toilet bowls. It's called a *closet auger*.

CLEARING FLOOR DRAINS

Sand, dirt or lint often clogs floor drains. Remove the strainer and scoop out as much sediment as you can. You may have to chip away the concrete around the strainer carefully to free it. Flush the drain with water. If pressure is needed, use a garden hose.

Occasional flushing of floor drains may prevent their clogging. Pour a pail of water down them every few months or so.

Roots growing through cracks or defective joints in outside drains or sewers sometimes clog them. You can clear the stoppage temporarily with a root-cutting tool. To prevent future trouble, the line should be dug up and relaid using sound pipe and making sure all joints are watertight.

LEAKY PIPES, TANKS

Water sometimes corrodes metal pipes. This usually occurs all along the pipe rather than at one point. An exception would be where pipes of dissimilar metals, such as copper and steel, come together. The solution is either water treatment to prevent corrosion or the use

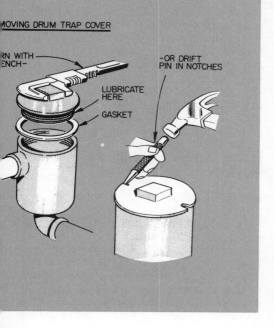

TURN WITH WRENCH

-OR DRIFT PIN IN NOTCHES

LUBRICATE HERE

GASKET

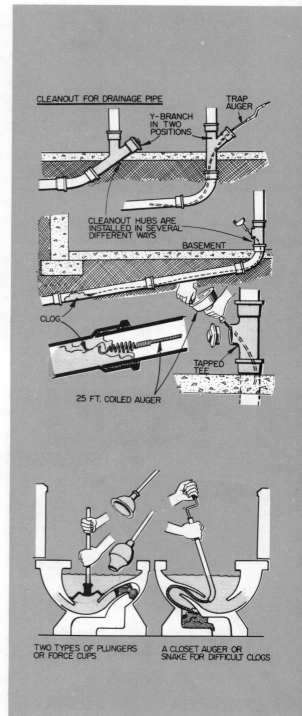

CLEANOUT FOR DRAINAGE PIPE

TRAP AUGER

Y-BRANCH IN TWO POSITIONS

CLEANOUT HUBS ARE INSTALLED IN SEVERAL DIFFERENT WAYS

BASEMENT

CLOG

TAPPED TEE

25 FT. COILED AUGER

of pipe materials that are more compatible with your water. Consult a local plumber.

Pipes split by hard freezing must be replaced.

A leak at a threaded connection often can be stopped by taking apart the fitting and making it up again with a good pipe joint material in the threads.

Small leaks can be repaired with special rubber clamp-patches made for that purpose. Treat these as emergency repairs. Later the section of pipe should be replaced.

A large leak in a pipe may call for cutting out the damaged section and installing a new piece of pipe. At least one union will be needed unless the leak is near the end of the pipe.

An emergency pipe repair can be made with plastic or rubber tubing. The tubing must be strong enough to take normal water pressure in the pipe. Slip it over the open ends of the pipe and clamp it with pipe clamps or several turns of wire.

Vibration sometimes breaks a soldered

TWO TYPES OF PLUNGERS OR FORCE CUPS

A CLOSET AUGER OR SNAKE FOR DIFFICULT CLOGS

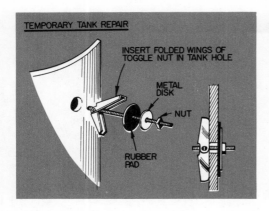

TEMPORARY TANK REPAIR

INSERT FOLDED WINGS OF TOGGLE NUT IN TANK HOLE

METAL DISK

NUT

RUBBER PAD

only one spot on the tank wall, the wall may be thin in other spots too. For this reason, any repair is temporary. The tank will undoubtedly have to be replaced before very long. Temporary repair lets you replace at the most convenient time, not under duress.

A tank leak can be temporarily repaired with a toggle bolt, rubber gasket and brass washer. You may have to drill out the hole to get the toggle bolt in. Draw the bolt up tight to compress the rubber gasket against the tank's side (see drawing).

FROZEN PIPES

In cold weather, water may freeze in unheated locations such as crawlspace or along outside walls. When water freezes it expands. Unless the pipe material can expand too, it may split. This is typical of iron and steel pipes. Copper pipe will expand some, but doesn't come back to its original size after freezing. Too many

joint in copper pipe causing a leak. If the joint is accessible, clean and resolder it. All water must be removed from the pipe before you can heat it to soldering temperature.

Leaks in tanks are usually caused by corrosion. Sometimes a safety valve may not open and the pressure developed will spring a leak. While a leak may show at

If torch is used on frozen pipe, do not get pipe so hot it can't be held by hand.

A safer way to thaw a frozen pipe: pour hot water over rags and wrap around pipe.

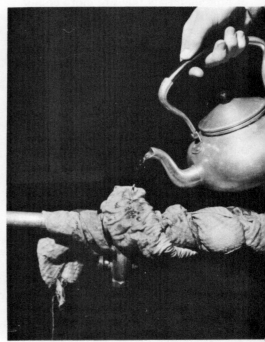

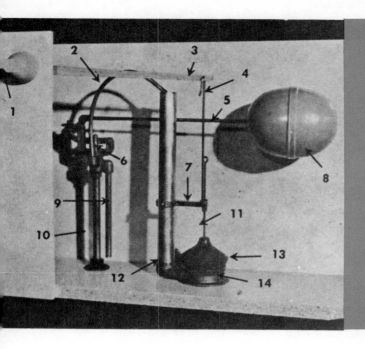

1. Trip handle
2. Refill tube
3. Trip lever
4. Upper lift wire
5. Float arm
6. Ball cock supply valve
7. Guide arm
8. Float
9. Discharge pipe
10. Supply pipe
11. Lower lift wire
12. Overflow pipe
13. Flush ball
14. Flush valve seat

freezings and it will split, too. While flexible plastic pipe will take lots of freezing and thawing, it's wise to keep it from freezing.

You can prevent freezing by insulating the pipe or by installing an electric heating cable to warm it. Wrap the cable around the pipe and cover it with insulation.

The best, safest method of thawing a frozen pipe is with an electric heating cable, because the whole length of pipe is thawed at once. Thawing with a torch can be dangerous. Steam generated can make the pipe blow up in your face. If you must use a torch, never get the pipe hotter than you can hold your hand on.

Thawing with hot water poured over rags wrapped around the pipe is safer than torching.

When flame-thawing a pipe, open a faucet and start thawing at that point, reducing the chance for dangerous pressure build-up.

TOILET TANK REPAIRS

Toilet tank trouble nearly always spells a leak of some kind. Either the flush valve is failing to seat or the float valve isn't stopping water from entering the tank once it's filled. Hundreds of gallons of water can slip through a toilet tank unnoticed. At a rate of 20 cents a thousand gallons for city water, several cents a day can disappear down the drain. It may not sound like much, but you'd need something like $3000 invested at 4 percent interest to stay even with the loss. The excess water runs into the overflow pipe and down into the toilet bowl. Gone.

Once in a while something else goes wrong with a toilet tank, but a few adjustments can usually fix it. When there's a leak you can usually hear water running, if only slightly.

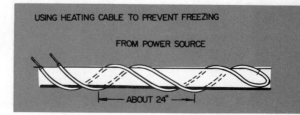

USING HEATING CABLE TO PREVENT FREEZING

FROM POWER SOURCE

ABOUT 24"

Easy way to tell if toilet tank is leaking is to place a piece of paper to the back of bowl. There's a leak if paper gets wet.

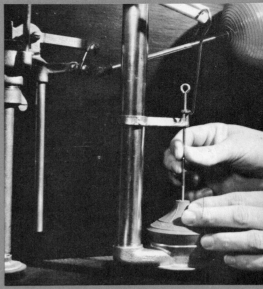

Installing a new flush ball and new lift wires takes care of a leak due to deterioration of flush ball or bent lift wires.

You can adjust water level by bending the float arm. Bend up to raise, down to lower level to ¾-in. below top of overflow pipe.

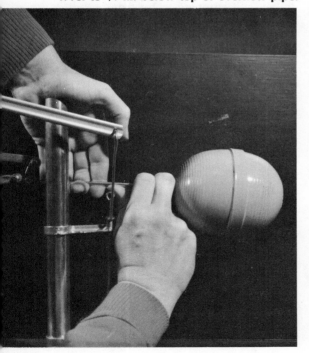

Here's a quick test that will help you find even a slight toilet tank leak in either the flush valve or float valve system. Touch a piece of paper to the back of the toilet bowl, above the waterline. If the paper gets wet there's a leak.

Take the flush valve system first. There are different types of flush valves in common use. The Douglas-type valve with flush ball is the most common. The newest toilets are equipped with either a brass or plastic Douglas valve and rubber flapper tilt, or bucket stopper. Other combinations are flapper with china seat and flush ball with china seat.

The older flush ball types are operated with upper and lower lift wires guided by arms. The flappers are operated with stainless steel link chains. As long as you keep any type of flush valve clean and well adjusted, you shouldn't have any trouble with it. If a ball or flapper gets coated with minerals or scum, if the rubber parts are defective, if the seat is irregular or if the lift wires are bent, the valve may leak. A ball or flapper that's misshapen, worn out, or one that has lost its elasticity and fails to drop tightly into the hollowed valve seat should be re-

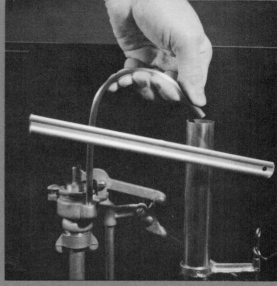

If float valve still leaks and float arm is free, remove inlet valve plunger and install a new washer and place new packing on it.

Bend refill tube down so its water spews into end of overflow pipe. Don't however, get it below level of water in the tank.

placed with a new one.

RENEWING A TANK BALL

To replace a tank ball, first turn off the water supply. Usually there's a stop valve for this purpose under the toilet tank. You can also cut the water by propping a stick under the float arm inside the toilet tank. Then unscrew the flush ball from the lower lift wire. Attach a new ball the same diameter as the old one. If the threads are corroded, you may have to snip off the lower lift wire and replace it. No problem. You can buy new wires most anywhere that plumbing supplies are sold. Consider replacing both of the lift wires anyway. It's cheap insurance. Get brass ones if you can. If you must re-use the old wires, inspect carefully to see that they're not bent. Before you put in the new flush ball, scour the valve seat with fine steel wool or No. 500 wet-or-dry abrasive paper. Get a smooth, uniform bearing for the stopper.

After you install the new tank ball, check to make sure it's centered directly over the seat. Adjust the guide arm if you

Tightening nut behind trip lever is about all the service it needs. Use adjustable open-end wrench on left hand threaded nut.

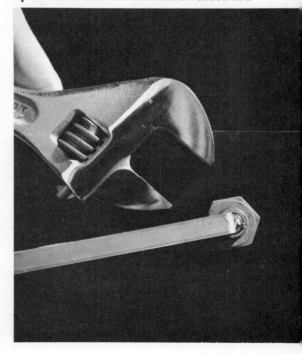

103

You can dry a sweating porcelain toilet tank by cementing insulation to inside.

A pipe patch can be a big lifesaver. Keep one on hand in case you find a leaky pipe.

need to. Some can be made longer or shorter as well as moved sidewise around the overflow pipe.

You can buy a new tank ball guide with a longer slide that reduces the amount of side-play in the flush ball. This may solve your problem. A new stopper with a bottom projection may solve a misalignment problem.

To equip your toilet tank with a rubber flapper flush valve, remove the flush ball, lift wires and ball guide. Slip the flapper onto the overflow pipe. Adjust it to drop neatly into place on the valve seat. Then hook up the trip chain to the right length. If the overflow-to-valve-seat distance on your Douglas valve isn't standard, a flapper may not work right.

Plastic flush valves can be cracked, letting water pass. Brass valves can be scored from past leakage. Replacement of the whole assembly is the only sure cure for these maladies.

FLOAT VALVES

Float valves are the devices that shut off the water to the toilet tank when the level reaches a certain height. They're of two types: vacuum-breaker and nonvacuum-breaker. Vacuum-breaker types have their bonnets and air inlet ports at least an inch above the overflow level of the tank. They're much the preferable of the two because they prevent toilet tank water from being accidentally siphoned back into the house water supply and possibly contaminating it.

Nonvacuum-breaker valves are completely submerged in the tank water.

If your tank leaks at the float valve rather than the flush valve, check to see that the float isn't hanging up against the trip lever or the side of the tank, thus failing to close the intake cock. Bending the float arm will correct this. Unscrew the float and shake it to see whether any water has leaked inside. If it has, replace the float. Some tank floats are plastic, eliminating the corrosion problem of copper floats.

Sometimes a readjustment of the water level in the tank is what's needed to stop the leaking or to provide enough water for a good flush. This is done by bending the metal rod attached to the tank float. When the rod is bent upward, the water

It is much cheaper to repair worn, leaky faucets and valves than to buy new ones.

will rise higher in the tank. When the rod is bent downward, the water level will be lowered. The water level in most toilet tanks should be 3/4 inch below the top of the overflow pipe.

To fix a balky float valve, take it apart. To do this on the common type, turn off the water supply. Remove the screws that hold the float arm levers and lift them out. The stop cock will then lift out. Pull off the old washer with a pair of pliers and install a new one. Usually this is sufficient to stop the leaking. Sometimes the valve seat needs replacing, too. Then it may pay to replace the entire float valve assembly. Most hardware stores carry them in stock. The cost isn't much.

REFILL TUBE

While you're at it, check the refill tube to see that it isn't plugged and that it's discharging into the overflow pipe. But make sure it isn't bent down into the pipe below the water level. If it is it'll siphon water out of the tank continuously and you won't even know it.

If you have to hold the trip lever down until the toilet is through flushing, in order to get a complete flush, the trouble is probably in the lift wires. They may not be raising the tank ball high enough so that the force of the outrushing water can't pull it back down again. If you straighten and rebend the upper lift wire to shorten it, that should fix your trouble.

Rarely is the trip-lever the cause of toilet tank problems. There are two types in use on all toilets: single-acting and double-acting. The single-acting kind is more widely used today and seldom needs more than an occasional tightening of the lefthand threaded nut that holds it in the tank opening.

TOILET BOWL LEAKAGE

If a toilet bowl leaks around the bottom, it will have to be removed and resealed with a new bowl gasket. Follow this procedure:

1. Shut off the water and empty the tank and bowl by siphoning or sponging it out.

2. Disconnect the water pipes to the tank.

3. If the toilet and tank are a two-piece unit, disconnect the tank and bowl. Set the tank where it can't be damaged. Always handle the parts carefully. They're made of vitreous china or porcelain which is easily chipped or broken.

4. Remove the seat and cover from the bowl.

5. Carefully pry loose the bolt caps and remove the bolts holding the bowl to the floor flange. Jar the bowl enough to break the seal at the bottom. Lift the bowl off and set it upside down on a thick padding of newspapers.

6. Clean off all of the old sealing material from the bowl and floor flange. Place a new wax or rubber bowl gasket around the bowl horn and press it into place.

7. Set the bowl as described in the chapter on installing fixtures. Install the tank and connect the water pipes to it. It pays to install all new gaskets, after first cleaning the surfaces thoroughly.

8. Test for leaks by flushing a few times. Then bolt on the seat and cover.

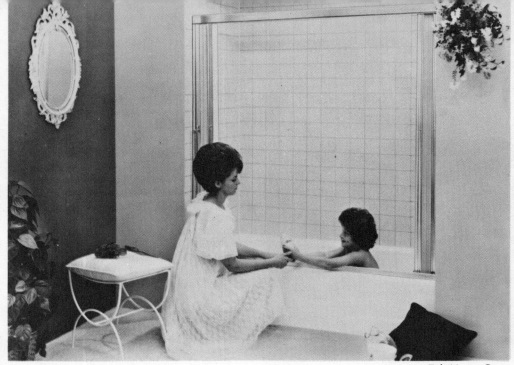

Tub-Master Corp.

Adding a modern tub-shower to your bathroom can add beauty and value to your home.

HOW TO MODERNIZE YOUR PLUMBING

If your house is showing its age, give it a fixture lift

You can improve an old home by modernizing its plumbing in two ways: (1) add new plumbing facilities; (2) replace the old fixtures with modern ones. New facilities can be such as a new bathroom, or half-bath, shower, double sink with garbage disposal, dishwasher, first-floor laundry, outdoor shower, etc. New fixtures are put in place of the old to make a house better looking, more comfortable to live in and more salable.

The biggest problem when adding new plumbing facilities to an existing house is hooking new piping into existing drain-waste-vent system. The old system is hidden behind the walls. You may have to cut into the walls to effect the hook-ups. Concealment of the new pipes can be a problem, too. You may find considerable carpentry necessary to hide your new drain and water supply pipes. The standard methods of running pipes in walls may not be as practical as building enclosures for the pipes to run in. To reach a second story from a basement or crawlspace, you can often route pipes unobtrusively through a closet. Then no enclosure is needed. An enclosure can take the form of a soffit, corner cabinet, room divider or closet.

Plan, if you can, for the new facility to use a wall with an existing facility on the other side. Then you can tap into pipes that are already there. If the new plumbing includes a toilet, the existing plumbing will have to include one, too. Otherwise you'll need to install a new main stack running into the main house drain. That can be a big job.

Horizontal runs of new pipe are easier to accommodate than vertical runs. You can go horizontally in a basement, crawl-space or attic without much trouble. To go vertically you may have to knock out wall material to expose the framing. Later, the finished wall will have to be replaced. If the room can benefit by a whole new wall surface after remodeling its fixtures, then don't worry about tearing it up. Hack away.

Concealed piping requires cutting through some studs and joists. The worst are the cuts for drain pipes because of their larger sizes. Water supply pipes are smaller and easy to run through framing members. They're usually run in the same spaces as drain pipes.

Remember that threaded pipes and fittings need more room in the wall than nonthreaded ones. The additional room is for clearance as you turn the fittings to tighten them. For this reason, plastic and copper are preferred pipe materials for remodeling. No-hub cast iron drain pip-

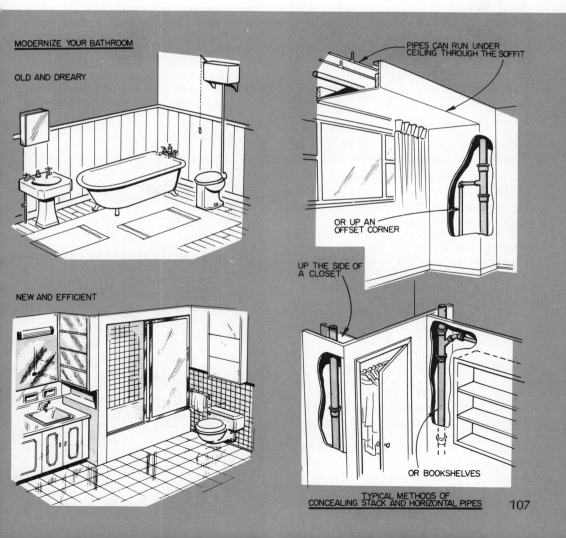

MODERNIZE YOUR BATHROOM

OLD AND DREARY

NEW AND EFFICIENT

PIPES CAN RUN UNDER CEILING THROUGH THE SOFFIT

OR UP AN OFFSET CORNER

UP THE SIDE OF A CLOSET

OR BOOKSHELVES

TYPICAL METHODS OF CONCEALING STACK AND HORIZONTAL PIPES

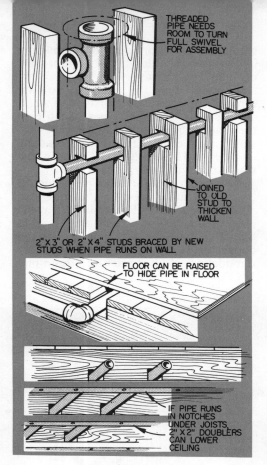

THREADED PIPE NEEDS ROOM TO TURN FULL SWIVEL FOR ASSEMBLY

JOINED TO OLD STUD TO THICKEN WALL

2" X 3" OR 2" X 4" STUDS BRACED BY NEW STUDS WHEN PIPE RUNS ON WALL

FLOOR CAN BE RAISED TO HIDE PIPE IN FLOOR

IF PIPE RUNS IN NOTCHES UNDER JOISTS 2" X 2" DOUBLERS CAN LOWER CEILING

Fitting pipes into walls.

ing is good, too. Try to avoid hub-and-spigot cast iron and threaded pipe.

If a partition isn't thick enough for your new pipes to go through, set 2×4 studs against the existing studs after the walls have been opened and the pipes installed. The pipes can be notched halfway into each set of studs, so as not to weaken either one too much.

In an outer wall with plaster directly over bricks, you'll have to add a new stud wall—2×4, 2×6 or even 2×8—for sufficient thickness to handle the pipes.

Cut notches in studs to pass your piping. Make parallel cuts with a saw and knock out the wood between with a plumber's chisel. Don't notch the lower half of a stud deeper than one-third its width without reinforcing it later with a steel strap or furring strip. Never notch the lower half of a stud deeper than two-thirds. You may notch the upper half to

108

one-half its depth without reinforcing it. This is in a nonbearing partition wall only.

STACK CONNECTIONS

If you are adding a fixture during remodeling and no provision was made when the soil stack was built, you'll have to break into either the existing stack or the house drain. Break out a section of pipe and remove it. If it is cast iron, remove the whole piece of pipe broken into. If it is copper or plastic, saw squarely across the run, removing no more pipe than you have to. Support the upper portion of a severed stack with some sort of bracing to prevent sagging or falling. Install a tee or wye fitting for the new branch drain. For a cast iron stack use a sission joint, as shown in a drawing in this chapter. For a plastic or copper stack install a collar as shown. To make it, saw a coupling in half. A special saddle fitting is available for solvent-welding to a plastic stack. After welding, a hole is bored through the stack behind the

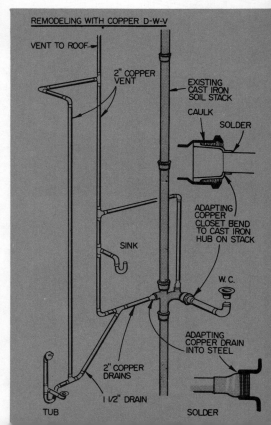

REMODELING WITH COPPER D-W-V

VENT TO ROOF

2" COPPER VENT

EXISTING CAST IRON SOIL STACK

CAULK

SOLDER

SINK

ADAPTING COPPER CLOSET BEND TO CAST IRON HUB ON STACK

W. C.

ADAPTING COPPER DRAIN INTO STEEL

2" COPPER DRAINS

1 1/2" DRAIN

TUB

SOLDER

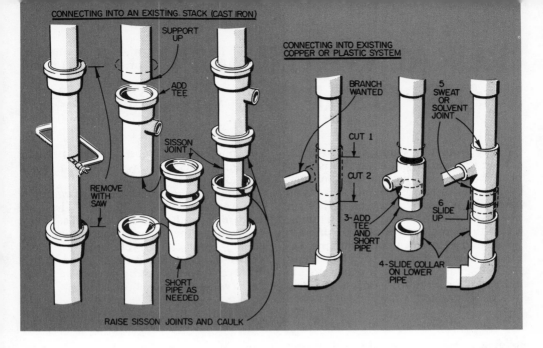

SUPPORT UP

ADD TEE

SISSON JOINT

REMOVE WITH SAW

SHORT PIPE AS NEEDED

RAISE SISSON JOINTS AND CAULK

CONNECTING INTO EXISTING COPPER OR PLASTIC SYSTEM

BRANCH WANTED

CUT 1

CUT 2

3-ADD TEE AND SHORT PIPE

4-SLIDE COLLAR ON LOWER PIPE

5 SWEAT OR SOLVENT JOINT

6 SLIDE UP

fitting. The fitting can be placed anywhere on the stack without cutting the pipe.

NEW FIXTURES

Ordinarily there are no special problems in replacing an old fixture with a new one. The water supply and DWV piping are already there. If the new fixture is to be centered pretty nearly where the old one was, you're in. If it is to be moved much, you may have to remodel the DWV fixture branch to stay within code limitations on maximum length of wet-vented drain run. The water supply pipes will have to be extended, too.

When you select your new fixtures, remember that there are two qualities of plumbing: the best and the cheapest. Take your pick. The difference in cost may be only one-third. The quality of your fixtures should match the quality of your house.

When you modernize, get a modern size fixture, too. Small bathtubs are out. A size 60 inches long 30 or 32 inches wide and 16 inches high is now standard. Anything less defeats modernization.

Fixture trim should be solid brass with heavy nickel or chrome plating.

Information on toilet types is given in another section. The best toilet seats are solid plastic. Plastic-coated wood is in second place and painted wood or hollow plastic is rated as cheap. Toilet lids with iron hinges soon look bad when the hinges rust. Avoid iron hinges.

Stay away from painted metal shower cabinets. They soon rust out. You're better off to buy a terrazzo or composition shower base. Then build your own shower cabinet out of one of the waterproof vinyl-surfaced wallboards or of ½-inch plywood, exterior type C-C plugged. You can use ceramic tile with either material or paint the plywood with epoxy paint.

If any of your present fixtures have dangerous submerged inlets — faucet outlets below the flood rim of the basin — replace them. The risk of contaminated water being drawn into house plumbing where someone may later drink it isn't worth taking.

A good fixture modernization might be the installation of a better toilet with color-matched lavatory and tub. Another might be to replace a standing lavatory with one that has a built-in cabinet with storage space. You can buy the cabinet unit complete with bowl or make the cabinet portion and add the bowl to it.

Plan your remodeling carefully. The better you plan, the less work it will be, you'll find.

RURAL WATER SUPPLY AND DISPOSAL

Water problems you will have if you build outside of cities

A way from cities and suburbs there are no municipalities to furnish water and sewer services to homeowners. Every man for himself. While there is no water bill, no sewer assessment, it's not the most economical way of doing things. A few thousand dollars can be buried in the ground before you have running water and sewer. Plan on spending about $1000 or more to create a water system and $1500 or more for a decent disposal system. Then there's the electricity used for pumping well water and the periodic cleaning necessary with most disposal systems.

The most common source of water for a rural home is the drilled well. There are also dug wells and driven wells. Water can sometimes be obtained from lakes and streams but rarely are they clean enough. Lacking a more dependable source, a spring can be developed to furnish water for household needs.

Rain water that seeps deep into the

Building a rural sewage disposal system is lots of work. Leave it to professionals.

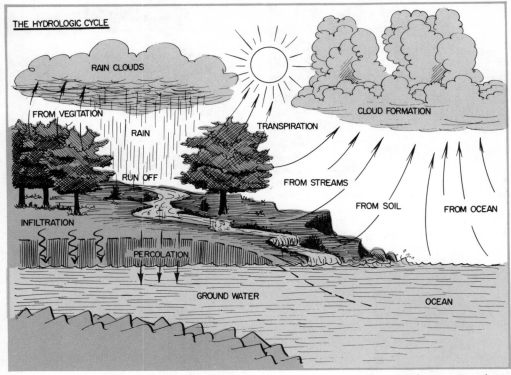

THE HYDROLOGIC CYCLE

RAIN CLOUDS

FROM VEGITATION

RAIN

RUN OFF

INFILTRATION

PERCOLATION

GROUND WATER

TRANSPIRATION

CLOUD FORMATION

FROM STREAMS

FROM SOIL

FROM OCEAN

OCEAN

Water Conditioning Foundation

ground is usually the best, most reliable water source. This can be tapped by drilling a well down into it. The well depth needed to reach water varies a lot with locality. Wells 100 feet deep are common. Some go to 300 feet. In parts of the Southwest it's possible to bore down thousands of feet and not find water.

JOB FOR A PRO

The most practical way to get any but the shallowest well is to have it drilled by a professional. He will use one of a number of methods for augering through the earth. A popular one is called the *percussion* method. With it, a machine raises a heavy steel bit and drops it repeatedly to chew out a hole deep into the ground. Lengths of 5- to 8-inch steel casing pipe are fastened together with huge threaded couplings and driven down until they reach a layer of rock. The casing not only keeps your well from collapsing, it pre-

vents well pollution by contaminated surface water. In rock, the drilled hole can't collapse and often is not lined. In sandy soil a well may have to be cased all the way down to prevent collapse. Most well-drillers prefer to quote you a per-foot price and let you gamble on how deep the well and casing must go.

Locate the well to keep the pipe run between well and house as short as practical. The well should not be put in the basement because if deepening is ever needed, the driller's rig can't work there.

The in-way to cap off a drilled well is with a what's called a *pitless adapter*. The well casing is brought above ground and capped to keep dirt and vermin out. Below the frost line a side-tapping in the casing permits the water pipe to leave the casing and run in a trench to your house. The pitless adapter's brass fitting fits tightly into a hole drilled through the side of the well casing pipe. The hookup is such that, should the water pipes ever have to be pulled from the well, it can be

A well-driller's rig can bore down hundreds of feet, lining the well with pipe.

Rock-cutting bit lowers into well casing on a cable and chisels away the earth.

A slim submersible pump is screwed onto the well pipe and lowered into the casing.

Last length of pipe is a stub that serves as a handle; hooks onto a pitless adapter.

done merely by removing the cap and lifting them out. No digging is involved. What's more, no well pit is needed. The necessary pump accessories all go in the basement or utility room.

DO-IT-YOURSELF WELLS

If you want to make the well yourself, go ahead. It's no fun, though. In loose earth, sand or gravel you can make what is called a driven well. A hardened steel well point with mesh-screened sides is screwed onto galvanized steel pipe and driven with a sledge. A short piece of

pipe coupled to the top takes the battering. In soft ground you can drive a well 100 feet or so. If your driven well pipe misses all buried rocks and keeps going until you hit water, consider yourself lucky.

A problem with driven wells is that water often hides just below a layer of rock. Your well point can get down to the rock but not through to the water. All you can do is cry a lot.

Shallow wells of 15 to 30 feet can be dug by hand, if you have nothing better to do. Rain water filters through the ground and collects in them. They should

Steel casing pipe is driven down to rock formation to keep sides from caving in.

Pitless adapter fastens through a hole in well casing and connects with house pipe.

be lined down to the water table to prevent infiltration of surface water. Besides the hard work, the chief problem with dug wells is that, lining or not, they aren't deep enough to avoid contamination. And, they may not supply water all year around. Surface water has a way of drying up.

Ask around. See where other people in the area get their water. What problems do they have? Is it good water? Drink some. Talk to well-drillers. Are they certain of finding water for you? Get a rough idea of how much it usually costs.

Wherever you get water, have it tested for potability by your local health department. Most furnish sterile bottles for water sampling. Good looking, clear water may contain bacteria, nematodes and other unseen guys wearing black hats. You won't want to drink them. If the water source could be infiltrated by surface water, have it checked at regular intervals.

PUMPING WELL WATER

To get city water service from a country well, you'll need some kind of pumping system. It consists of an electric motor, pump, water-air pressure tank, pressure switch and a device for injecting air into the pressure tank. The air-injection keeps the pressure tank from becoming waterlogged.

Well pumps are efficient and reliable. Your pump must have steam enough,

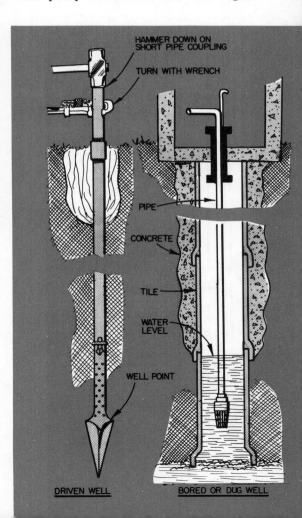

HAMMER DOWN ON SHORT PIPE COUPLING

TURN WITH WRENCH

PIPE

CONCRETE

TILE

WATER LEVEL

WELL POINT

DRIVEN WELL

BORED OR DUG WELL

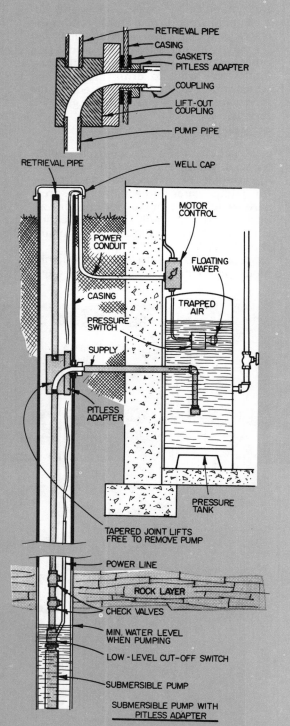

RETRIEVAL PIPE
CASING
GASKETS
PITLESS ADAPTER
COUPLING
LIFT-OUT COUPLING
PUMP PIPE

RETRIEVAL PIPE
WELL CAP
MOTOR CONTROL
POWER CONDUIT
FLOATING WAFER
CASING
TRAPPED AIR
PRESSURE SWITCH
SUPPLY
PITLESS ADAPTER
PRESSURE TANK
TAPERED JOINT LIFTS FREE TO REMOVE PUMP

POWER LINE
ROCK LAYER
CHECK VALVES
MIN. WATER LEVEL WHEN PUMPING
LOW-LEVEL CUT-OFF SWITCH
SUBMERSIBLE PUMP

SUBMERSIBLE PUMP WITH
PITLESS ADAPTER

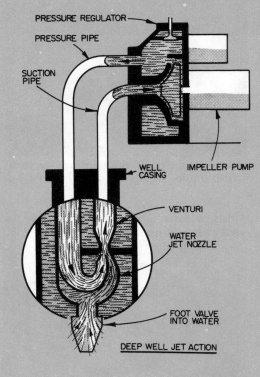

PRESSURE REGULATOR
PRESSURE PIPE
SUCTION PIPE
IMPELLER PUMP
WELL CASING
VENTURI
WATER JET NOZZLE
FOOT VALVE INTO WATER

DEEP WELL JET ACTION

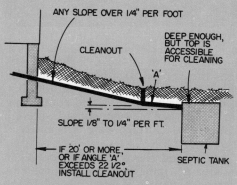

PREFERRED SEWER POSITION

ANY SLOPE OVER 1/4" PER FOOT
DEEP ENOUGH, BUT TOP IS ACCESSIBLE FOR CLEANING
CLEANOUT
'A'
SLOPE 1/8" TO 1/4" PER FT.
IF 20' OR MORE, OR IF ANGLE 'A' EXCEEDS 22 1/2° INSTALL CLEANOUT
SEPTIC TANK

however, to deliver all the water you'll need to draw at one time. If the capacity of a well is too small, you'll have to build a storage tank and pump water into it slowly. Water can then be withdrawn as rapidly as needed.

To figure well and pump capacity, multiply the number of fixtures in your house by 60. That's it in gallons per hour. Consider each hose outlet, automatic washer, laundry tub, dishwasher, etc., as a fixture. The recommended minimum pump capacity for residences is 540 gallons per hour.

Air compressed in the pump's pressure tank keeps your house water under pressure so it flows out when you open the tap. When the tank pressure drops to a predetermined level, say 30 psi., a pressure switch starts the pump and refills the tank with water, compressing air at the top of it. At peak pressure, say 50 psi., the switch cuts off and the pump stops. In the newest systems a plastic wafer floats on top of the water in the tank to reduce dissolving of tank air in water. This prevents waterlogging. It also permits prepressurizing of the tank for increased storage capacity.

Water in a well usually comes way up above the bottom. When your well driller completes his job, he'll be able to tell you how deep the water holds at full pumping capacity. The pumping equipment you buy should be based on that depth.

If, under the well driller's rapid bailing, the water level stays at 25 feet or less, you can get excellent service from a shallow-well pump. Most of these suck

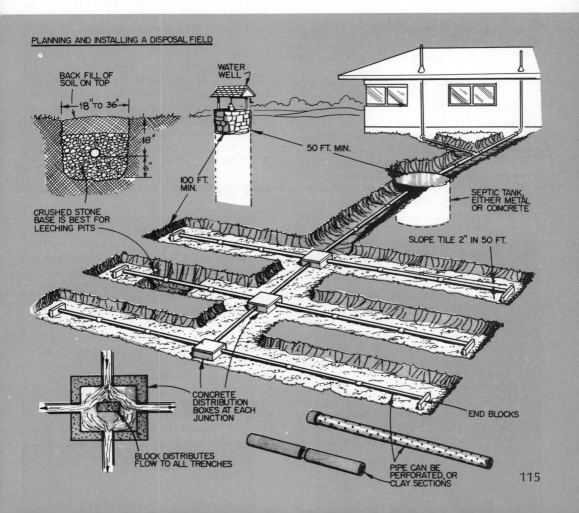

PLANNING AND INSTALLING A DISPOSAL FIELD

BACK FILL OF SOIL ON TOP

18" TO 36"

18"

6"

WATER WELL

50 FT. MIN.

100 FT. MIN.

CRUSHED STONE BASE IS BEST FOR LEECHING PITS

SEPTIC TANK, EITHER METAL OR CONCRETE

SLOPE TILE 2" IN 50 FT.

CONCRETE DISTRIBUTION BOXES AT EACH JUNCTION

BLOCK DISTRIBUTES FLOW TO ALL TRENCHES

END BLOCKS

PIPE CAN BE PERFORATED, OR CLAY SECTIONS

115

Painting dope on clay spigot, and the hub inside, is an easy way to make joints.

water up out of the well like milk from a straw. A small shallow-well pump can deliver lots of water because of the short lift.

A lower water level under bailing calls for a deep-well pump. These push the water up because suction isn't effective beyond 25 feet. Pumps can be submersible, jet or piston pumps. Most of those installed these days are submersibles. Pumps and motor are installed below the lowest well water level. They pump more water, are quieter and more trouble-free than the others. Though they cost more, only one pipe down the well is required. A submersible pump that's designed for it will push water up from nearly 500 feet, if need be. Well diameter should be 4 or 5 inches. If the water is sandy, forget a submersible. Sand will wear it out in no time. Use a jet pump instead.

A jet pump is effective to 250 foot depths and more. Two pipes are needed in the well. Water is pumped down one pipe, around a hairpin curve and back up through a venturi. The area beyond the venturi is open to well water and water is drawn in from the well. The flow carries it along up another larger pipe close enough to the surface that it can be sucked the rest of the way up. Meanwhile, back at the pump, excess water is

diverted that-a-way into the pressure tank where it's available for your house plumbing system.

With either a submersible or jet pump, when the pump shuts off, a foot valve at the bottom of the well piping closes, holding tank pressure in the entire system.

A shallow-well jet pump has its jet mounted on the pump. Only one pipe enters the well, which can be as small as 2 inches in diameter. A deep-well jet pump has its jet down in the well at about water level. Two pipes enter the well, which should be 4 inch diameter. In either case a pipe with foot valve extends below the lowest possible water level under continuous pumping.

A piston pump is used for shallow wells only. It works like an old hand pump that's been motorized. A mechanical lifting action brings up the water.

Water depth influences pump capacity. For instance a ½-hp submersible will provide 650 gallons per minute from a 20-foot depth. At a 160-foot depth it takes 1½ hp to give the same capacity. Jet pumps are more efficient at moderate depths. Submersibles are more efficient deeper. The jets are losing out because they're prone to stoppage by anything in the water large enough to lodge in the jet and plug it. Pipe scale can do that.

PIPE TYPES

The most practical kind of pipe to lower down a well is plastic. Use either flexible polyethylene or rigid PVC or ABS. Threaded joints or solvent-welded joints on the rigid pipes are okay. If the water isn't down too far, say not below 40 feet, you can haul the piping out by hand if pump problems develop. You can use 125-pound-test plastic pipe down to 200 feet with a jet and to 150 feet with a submersible pump. More economical 80-pound-test pipe may be used for jet pump installations to 100 feet deep. With poly pipe use double clamps on every fitting that carries the weight of the pipes and pump or jet. This prevents pull-off of fittings at depths to 150 feet. Below the maximum depths you'd better use steel

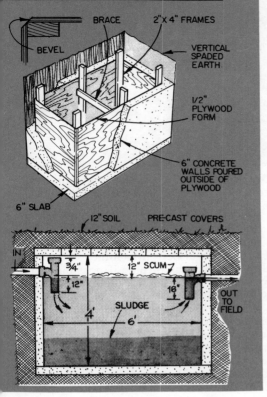

BRACE

2" x 4" FRAMES

BEVEL

VERTICAL
SPADED
EARTH.

1/2"
PLYWOOD
FORM

6" CONCRETE
WALLS POURED
OUTSIDE OF
PLYWOOD

6" SLAB

12" SOIL   PRE-CAST COVERS

IN

3/4"

12" SCUM

12"

16"

OUT
TO
FIELD

SLUDGE

4'

6'

SOIL PERCOLATION TEST

TIME REQUIRED FOR WATER TO FALL 1 IN.	FEET OF PIPE PER PERSON
1 Minute	12
2 Minutes	15
5 Minutes	20
10 Minutes	30
30 Minutes	60
60 Minutes	80
Over 60 Minutes	Unsuitable

SEPTIC SYSTEM

Where there is no city sewer, a sewage disposal system is needed to get rid of wastes after they leave the house. You can't run them out on the ground. That's forbidden. Neither can you dump them into a stream or lake. You must treat all sewage before disposing of it.

The simplest, most basic form of handling house sewage is the septic system. It comprises a sewer line, a septic tank, distribution lines and a seepage field. Rain water, ground water and nonseptic basement drainage should be kept out of the septic tank. One thing is sure about any septic system: it'll have to be replaced someday. Consider it as a way out of the sewage disposal problem, no more.

pipe. A plastic pipe submersible pump installation needs a safety line to prevent loss of the pump if the pipe should snap under starting torque. A submersible installation needs a low-level cut-off and a lightning arrester to protect the pump.

Well pumps operate on 110- or 220-volt electricity. Best is 220, if you have it. Run a fused electric line to a switch box by the pressure tank. Then wire the pump's switch to that. If you have circuit-breakers instead of fuses, you can eliminate the switch box.

Well water runs around under the ground for many years picking up some of the soluble minerals it traverses. These can include calcium, magnesium, iron, manganese and sulfer. Water treatment may be necessary to remove unwanted impurities and make the water better for drinking and washing. Have your water tested for mineral impurities. The necessary treatment units can be installed based on the tests.

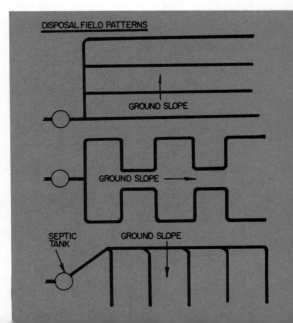

DISPOSAL FIELD PATTERNS

GROUND SLOPE

GROUND SLOPE

SEPTIC
TANK

GROUND SLOPE

To install a septic system, have backhoe dig for septic tank and disposal field.

Lowering a precast septic tank into the ground is done by the precaster's truck.

The entire septic tank and field installation usually need a building permit and the approval of your local health department. Get your health officer's advice before you begin. Locate the drain field on the downslope from your house but not where water collects. It should be at least 5 feet from the house and at least 50 feet from any well.

First build the sewer line leading from house to septic tank. It must be an enclosed pipeline to keep sewage in and roots out. A sewer line should be at least 4 inches in diameter and slope 1/8 to 1/4 inch per foot away from the house. Though the intervening sewer pipes may slope somewhat more without problem, the 10 feet of line just before the septic tank shouldn't slope more than 1/4 inch per foot. Install sewer pipes below the frost line but not too deep for efficient operation.

SEWER PIPE TYPES

Sewers may be made of cast iron, concrete, vitrified clay, cement-asbestos, pitch-fiber or underground plastic pipe. Check your code for the types you may use. Plastic is the lightest and easiest to work with. The rigid pipes—cast iron, concrete, cement-asbestos and vitrified clay—are stronger, but they are harder to work with and cost more. You can use them under unpaved driveways and in rocky soil. Get them with self-joints. Cast iron pipe with no-hub or compression joints is hard to beat as a sewer line. Pitch-fiber pipe comes with easy-to-

install tapered joints. Plastic pipe uses simple-to-make solvent-welded joints. They're described in another section. Buy long lengths of pipe if possible (10 feet is not too long). This avoids having to make lots of joints.

A sewer line should be placed on solid trench bottom cut to the proper slope. If you dig too deep, fill and level with crushed stones or gravel. Dig out small depressions for the hubs or couplings of each pipe so that pipes are supported along their whole lengths. Join the pipes and backfill gently at first. Plastic and pitch-fiber pipes need a selected backfill free of rocks and lumps that could squeeze the pipe. The hard pipes don't. Avoid high and low points in your trench bottom. Maintain a steady slope all the way. Never bed a sewer tile on blocks.

Often the first 10 feet of sewer line that extends through the house foundation and across the excavated area next to it is built of cast iron pipe. Try to use a single 10-foot length. Then the chosen type of pipe is carried the rest of the way. Make the connection with an adapter.

Cutting clay, concrete and cast iron pipes is similar. Chisel a line around them, pounding the cold chisel harder each time around until the pipe breaks off. Pitch-fiber and plastic pipe can be sawed. Buy couplings, 1/4 bends, 1/8 bends, tees, wyes and adapter fittings to suit the job.

SEPTIC TANK

A septic tank is a large watertight settling tank that holds sewage while it

Drain tiles are placed on sloping gravel fill. Joints are spaced apart for seepage.

decomposes by bacterial action. Sludge collects on the tank bottom. Liquid effluent flows out to the distribution system. A septic tank can be made of asphalt-coated steel, redwood, concrete, concrete block, clay tile or brick. You can make your own or buy one ready-made. The bottom and top of made-in-place tanks should be cast concrete.

Most septic tanks these days are of precast concrete. They're delivered to the site by the precaster and lowered into a hole dug in the ground. Then they're leveled. Be sure to backfill the septic tank hole before it rains. An empty tank can float up out of the ground.

A septic tank should be located where it can be pumped out every 2 to 5 years as the sludge level builds up. In cold climates there should be two feet of cover over the tank. Otherwise set the tank's top 12 to 18 inches below grade.

A septic tank should be sized to suit your family. Two-bedroom homes need minimum 750-gallon tanks, says the U.S. Public Health Service. Three-bedroom homes need 900-gallon tanks and four-bedroom ones require 1000-gallon tanks. Garbage disposers, automatic washers and dishwashers are figured in.

Since the effluent coming out of a septic tank is only partially treated, it must not be discharged above ground or into a lake or stream. The usual method is to run it into a seepage field. There, some effluent seeps into the ground. Some evaporates into the air. A seepage field can be a series of gravel-filled trenches or a big pit filled with stones and covered

with a thin layer of earth. The field should be at least 100 feet away from a drilled well (50 in special cases), and 10 feet away from buildings and property lines.

PERCOLATION TEST

Your public health officer will probably ask for a *percolation* test. This involves digging some post holes in the ground where your septic field is to go. They should be dug to the depth where your seepage piping will be placed, often 18 inches. Saturate the holes with water and leave overnight. Fill them again and see how long it takes the water to fall 1 inch. From this your health officer can advise how much drain field will be needed. His requirements may be based on those shown in the table accompanying this chapter. For example, if it takes 30 minutes for the water in the percolation test to fall 1 inch, 60 feet of drainage will be required for each person.

You have a wide selection of materials for field tiles to distribute water around the septic field. All are 4-inches in diameter. Perforated plastic pipe and pitch-fiber pipe in 10-foot lengths are great. One-foot lengths of concrete or clay tile are fine, too. These are laid with small spaces between tiles to permit seepage. In that way a drain field line is different from a sewer line. It's not supposed to be watertight. In fact it had better not be.

The drain lines are set on a bed of gravel, protected from infiltration by asphalt felt and covered by more gravel. Gravel permits dispersal of effluent. Top cover should be with porous fill such as black dirt. A lawn or trees can be grown over a septic seepage field, no problem, as long as the water stays underground. Choose trees that like lots of moisture. They'll get it.

Septic field design depends so much on the type of soil and slope of land that you need individual consultation. The best place to get it is from a septic system contractor. He will do the excavating or the entire job, as you wish. Never try to hand-dig for a septic system. Take in laundry or something to earn the extra money. You'll be glad you did.

HELPFUL PRODUCTS FOR YOUR PLUMBING JOBS

Special items and tools that will make your work easier

Your plumbing dealer has on his shelves a number of products — some of them new, some old — that can help you to better, easier plumbing. Use them. Mostly, they don't cost much. But they're a heck of a lot of help. Browse around until you have seen them all.

A few turns of Teflon packing prevents leaks in shower heads, garden hose fittings, drain slip joints, in faucets and valves.

A close-quarter tubing cutter, which rotates by hand or screwdriver, works where there isn't room to swing ordinary cutter.

Applying penetrating oil helps loosen a corroded pipe joint, fixture screw or a slip-nut. Allow time for oil penetration.

Teflon tape takes the place of messy pipe dopes. Wind it, stretching so the threads show through tape, and assemble as usual.

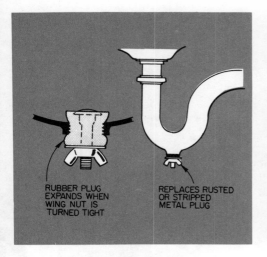

RUBBER PLUG
EXPANDS WHEN
WING NUT IS
TURNED TIGHT

REPLACES RUSTED
OR STRIPPED
METAL PLUG

Rubber clean-out plug.

Above right, the quick, easy way to tap water out of any pipe is with a saddle-tee clamped over a hole drilled in pipe.

At right, be careful when you use liquid drain opener. Pour it down the toilet or sink, as directed, and watch it go to work.

Self-tapping repair plug with rubber washer can save a tank or boiler. Sometimes a hole must be drilled to get them started.

Many plumbing products come packaged on cards, like this automatic valve used to bleed trapped air from hot water system.

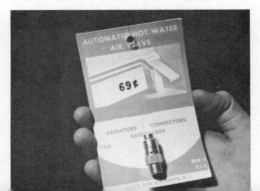

HOW TO PROCEED ON A HOME ADDITION

Plan your plumbing carefully for

maximum economy and efficiency

For radiant-hot-water-heating home, nail additional copper pipes to plaster back-up material spaced according to heating plans.

Heating and plumbing for a home addition is a big job. But if you know what you're doing, you can handle it all right. Tackle the jobs in this order: (1) heating, (2) rough plumbing (piping) and (3) finish plumbing. The heating is installed first because it's the least flexible. Even more flexible than plumbing is electrical wiring. Do that only after the rough plumbing is in.

You may either extend pipes or ducts from your present heating system into the addition or heat it with a separate system. The separate system can be as simple as wall or floor furnaces or, for tops in comfort, a complete central heating system. With a separate system you'll need to allow space in the addition for the heating plant and any chimney or venting required.

If you try to tap heat from the existing house system, be sure it's big enough to handle the load. Not many are. Actually they shouldn't be.

To know how much heat you need, you'll have to calculate the heat loss of

your addition. There are several methods of figuring heat loss, some more detailed than others. The Montgomery Ward general catalog has a section telling simply and clearly how to make the calculations. Your heating equipment dealer will be glad to figure it for you if you take him a set of plans. Any heat loss calculation is based on design temperature. This varies across the nation.

CHIMNEY

A new heating plant usually means a new chimney. Instead of building one all the way up from the basement, use a prepackaged chimney. It comes in sections that are screwed together. Such a chimney is fully insulated and is supported by the house framing. Only a small space is needed for it. Roof flashing and chimney cap are a part of the package. Your heating plant is vented into the lower end of the chimney once it's installed.

Fuel lines to the heating plant should

be run much like plumbing. Copper tube with flare fittings is suitable either for oil or gas.

The biggest job in installing the heating system is making the duct or pipe runs around to the rooms. You may have to cut holes here and there. Less cutting is involved with forced-hot-water heat than with forced-air. This is because water pipes are much smaller than air ducts.

You'll need some authorative advice on number and size of ducts to serve each room. The same with water pipes and radiators. Your heating equipment dealer is the best source. Have him help you plan the entire distribution system in return for your parts order.

In general, make your duct runs between joists when you can. Don't avoid across-joist runs like the plague, though. There's nothing wrong with them. Rectangular ducts are made in short sections to fit between wall studs. Round pipes come in longer lengths and are easier to work with. Fewer fittings are needed. Special fittings will get you from round pipe to duct for an up-the-wall run. Every duct run starts with a take-off at the furnace plenum. Then comes the duct itself, then elbow to bend it where you want it to go. Boots are made for connecting ducts to registers. A diffuser is installed in the end of each boot to beautify it and to spread out the warm air.

Hot water heating pipes are so small they can be run across or between ducts, little matter. They slip through holes bored up into the wall above. There they are connected to radiators.

Whether you use air or water heat, be sure to include some means of flow-adjustment in every run. This will permit balancing the system when it's finished.

The heating plant should come with instructions on hooking up its controls. Usually it is merely a matter of running wires between the different controls and wiring them up as shown. If you run into anything that's not clear, ask your dealer.

You should provide an electrical switch for shutting off power to the heating plant for use when you work on it, and turn-off during the summer.

Why build a masonry chimney for your addition when you can assemble a pre-packaged chimney, supporting it on the house framing.

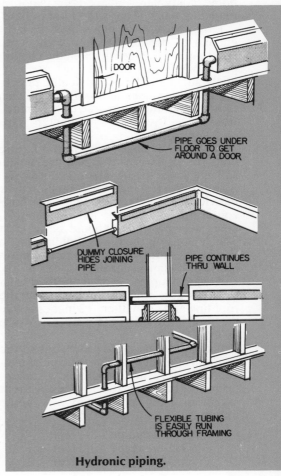

PIPE GOES UNDER FLOOR TO GET AROUND A DOOR

DUMMY CLOSURE HIDES JOINING PIPE

PIPE CONTINUES THRU WALL

FLEXIBLE TUBING IS EASILY RUN THROUGH FRAMING

Hydronic piping.

PLUMBING

The big item in plumbing is the 3- or 4-inch soil stack. It's expensive. The soil stack reaches from the house drain all the way up through the roof. Try to place all your fixtures as close to the stack as possible, within reason, or course.

Once your addition's plans are made, shop around among plumbing supply dealers to find one who's helpful. You'll need more information than we can give you here. The most help is someone who can answer your questions. Many dealers will also help you figure the materials needed to do a job. Others are plumber-oriented. They discourage do-it-yourself business. Try to find one that doesn't. Some dealers even offer discounts to home handymen who are doing major remodeling work. That's the kind to have.

Drain-waste-vent pipes must be large enough to handle the flow without restricting it. The sizes given in the chapter on DWV pipes and fittings are normally adequate. For complicated plumbing layouts, values are assigned to each fixture and their pipe sizes computed from these. (See tables in chapter "Tips on Running Pipes.")

To use the tables, add up the fixture units for all fixtures that will use a main soil stack. Size the stack accordingly. Size the waste pipes and vents for their individual fixtures. See what size horizontal run is needed to handle them. All pipes serving toilets should be at least 3-inch. Add up the total of all fixtures in the house addition and size the main house drain line accordingly.

Always increase roof vents to at least 3 inches at a point 12 inches below the roof line and carry them 12 inches above the roof line. This is to prevent closure by ice.

If there are no code provisions to the contrary, you can take advantage of wet-venting. Here are maximum distances from a fixture trap to the stack without reventing. They apply to any wet-vented fixture:

$$1\frac{1}{4}\text{-inch pipe} - 2\frac{1}{2} \text{ feet}$$
$$1\frac{1}{2}\text{-inch pipe} - 3\frac{1}{2} \text{ feet}$$
$$2\text{-inch pipe} - \quad 5 \text{ feet}$$

If any of your runs will be longer, you'll have to make revents for these fixtures. Remember, too, that wet-vents must not enter the soil stack below any toilet. Fixture drains that do, must be revented.

To make a 1½-inch revent, put a 1½-inch sanitary tee at the fixture stub-out behind the wall and continue the drain line as a vent run up above the fixture. Elbow it back and into the main stack above the point where the highest drain line is connected into the main stack.

While revent runs chop your stud walls full of notches, they sure beat drilling up through the roof. If long revent runs would be necessary, then go the secondary stack route. Usually this is a 1½- or 2-inch pipe leading up to the roof. A vent increaser below the roof line is required.

ASSEMBLING DWV

You'll nearly always find it easier to install a new main stack to serve the addition rather than to try tapping into an existing one in the old part of the house.

Start assembling a DWV system at the toilet drain. Cut all parts as necessary to fit. Sanitary tees may be ordered with 1½-inch side tappings for lavatory, shower or bathtub waste pipes. If you don't need one of the tappings, plug it.

Suspend the toilet drain parts from the joists or from nailed-on braces. Then work backward to them from the building's drain line in the wall or floor.

If the building drain and soil stack are different sizes or materials, you'll need an adapter or reducer between them.

Some provision should be made for rodding out every horizontal run of pipe, whether it's a 4-inch, 3-inch or a smaller waste pipe. This applies as well to a suspended horizontal drain, if one is necessary, between the soil stack and building drain. The usual method is through a clean-out opening at the far end of the pipe.

With the piping completed up to the toilet, build the soil stack up and out through the roof. Put in any branch tees for drains and revents as needed. Locate them so that all drains slope down ¼ inch per foot into the stack and all revents slope away from the stack. With cast iron you can use sanitary tees and invert them to get the proper slope as vent connections.

If the drainage portion of the soil stack has to be offset to clear a framing member, use ⅛-bends to do it. A ¼-bend may be used in the vent portion.

If a second floor toilet is to be plumbed in, build its drains first, then run the soil stack up to them from the first floor. This makes fitting easier.

Work out from the stack, building your branch waste and revent lines. If you use a secondary stack, build it up using the same procedure as for the main stack.

Stub out all drains so they can be fitted with fixture traps after the wall material has been installed. With steel pipe use a threaded nipple of any length loosely screwed in and capped until the fixture is installed. Then it can be replaced with what is needed for the fixtures. With plastic or copper, stub out at least an inch beyond the finished wall. You can saw the stub off the length later.

Framing that must be remodeled to accommodate drain pipes and fittings should be beefed up with cleats or headers. When a closet bend must cross a joist—never more than one—the joist can be cut short and its ends supported by headers as shown in the drawing in chapter "How to Modernize Your Plumbing."

WATER SUPPLY SYSTEM

Last of the rough plumbing is the water distribution hookup. Work from connections to the existing hot and cold water mains. If they're already overloaded, install new mains from the water heater to serve the addition.

Make your main line runs, then your branch runs. Stub out all pipes for fixtures in the wall if possible. Otherwise in the floor. Follow the rough-in dimensions illustrated.

Branch runs to fixtures should be carried to the point where they come through the wall to serve the fixture. At that point install tees. Going up from

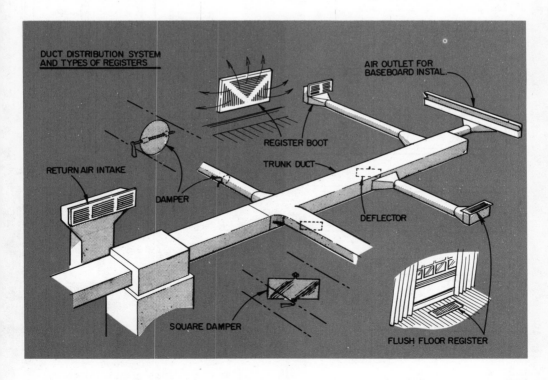

DUCT DISTRIBUTION SYSTEM AND TYPES OF REGISTERS

AIR OUTLET FOR BASEBOARD INSTAL.

REGISTER BOOT

RETURN AIR INTAKE

TRUNK DUCT

DAMPER

DEFLECTOR

SQUARE DAMPER

FLUSH FLOOR REGISTER

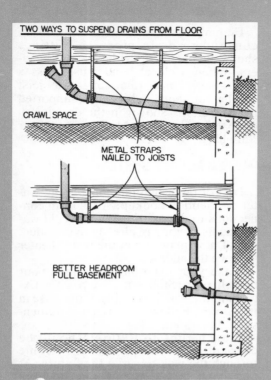

TWO WAYS TO SUSPEND DRAINS FROM FLOOR

CRAWL SPACE

METAL STRAPS
NAILED TO JOISTS

BETTER HEADROOM
FULL BASEMENT

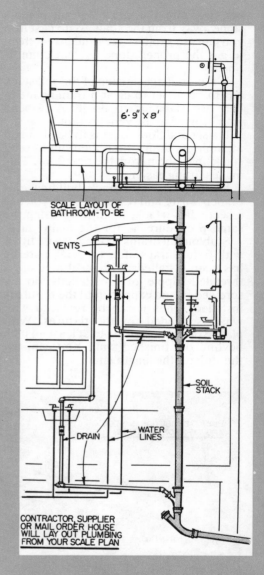

6'-9" X 8'

SCALE LAYOUT OF
BATHROOM-TO-BE

VENTS

SOIL
STACK

DRAIN

WATER
LINES

CONTRACTOR, SUPPLIER
OR MAIL ORDER HOUSE
WILL LAY OUT PLUMBING
FROM YOUR SCALE PLAN

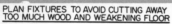

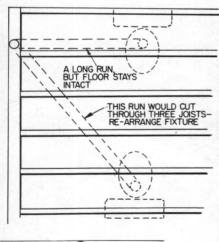

PLAN FIXTURES TO AVOID CUTTING AWAY
TOO MUCH WOOD AND WEAKENING FLOOR

A LONG RUN,
BUT FLOOR STAYS
INTACT

THIS RUN WOULD CUT
THROUGH THREE JOISTS—
RE-ARRANGE FIXTURE

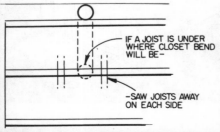

IF A JOIST IS UNDER
WHERE CLOSET BEND
WILL BE—

—SAW JOISTS AWAY
ON EACH SIDE

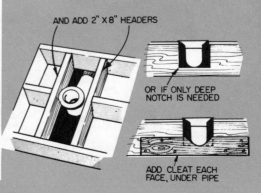

AND ADD 2" X 8" HEADERS

OR IF ONLY DEEP
NOTCH IS NEEDED

ADD CLEAT EACH
FACE, UNDER PIPE

INDEX

FIXTURE UNITS

Fixture	Unit Values	Pipe Size	Horizontal Pipe (¼″ per ft. slope)	Vertical Pipe
3-piece bathroom group— toilet, tub, lavatory	6	1½″	3	8
Toilet	4			
Bathtub	2	2″	6*	16
Shower	2			
Lavatory	1	3″	20**	30
Sink	2			
Laundry tub	2	4″	160	240
Floor drain	1			

* Waste only
** Not more than 2 tailets on horizontal line

MAX. HORIZONTAL LENGTH OF DRAIN PIPES
THAT ALSO ACT AS VENTS, OR 'WET VENTS'

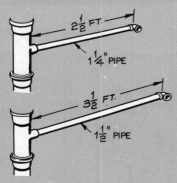

2½ FT. — 1¼″ PIPE

3½ FT. — 1½″ PIPE

5 FT. — 2″ PIPE

2½ FT. — 1¼″ PIPE

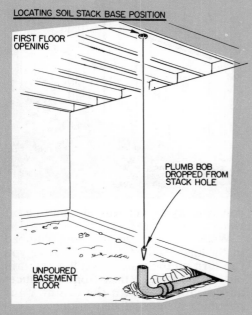

LOCATING SOIL STACK BASE POSITION

FIRST FLOOR OPENING

PLUMB BOB DROPPED FROM STACK HOLE

UNPOURED BASEMENT FLOOR

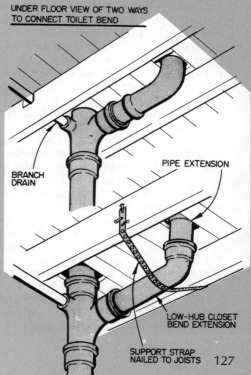

UNDER FLOOR VIEW OF TWO WAYS
TO CONNECT TOILET BEND

BRANCH DRAIN

PIPE EXTENSION

LOW-HUB CLOSET BEND EXTENSION

SUPPORT STRAP NAILED TO JOISTS

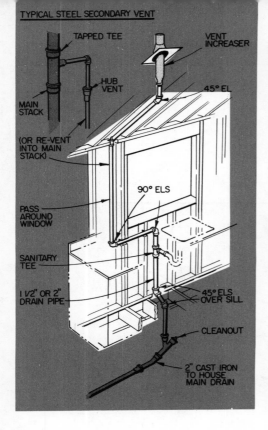

TAPPED TEE

VENT INCREASER

HUB VENT

MAIN STACK

45° EL

(OR RE-VENT INTO MAIN STACK)

90° ELS

PASS AROUND WINDOW

SANITARY TEE

1 1/2" OR 2" DRAIN PIPE

45° ELS OVER SILL

CLEANOUT

2" CAST IRON TO HOUSE MAIN DRAIN

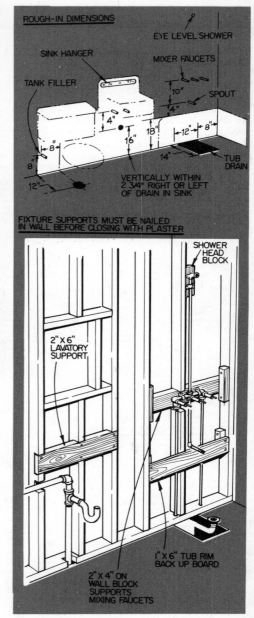

ROUGH-IN DIMENSIONS

EYE LEVEL SHOWER

SINK HANGER

MIXER FAUCETS

TANK FILLER

SPOUT

10"

4"

4"

18"

16"

12"

8"

8"

8"

14"

TUB DRAIN

12"

VERTICALLY WITHIN 2 3/4" RIGHT OR LEFT OF DRAIN IN SINK

FIXTURE SUPPORTS MUST BE NAILED IN WALL BEFORE CLOSING WITH PLASTER

SHOWER HEAD BLOCK

2" X 6" LAVATORY SUPPORT

1" X 6" TUB RIM BACK UP BOARD

2" X 4" ON WALL BLOCK SUPPORTS MIXING FAUCETS

each tee, put in a 12-inch-long length of capped pipe for an air cushion chamber. The branch of the tee gets a short length of pipe that should reach out an inch beyond the finished wall surface. On copper or plastic, come out farther. Cut off the excess later. Cap the nipple until you're ready to install the fixture. This will keep things clean inside. Capping is necessary for testing too.

As you install branch water supply runs to the fixtures, remember that the cold water is always the one on the right side as you face the wall. The hot water should be on the left. Even pros sometimes go wrong here.

Selective control of your water supply system is a good thing to have. One valve at the meter shuts off all water in the house. A valve at the hot water heater shuts off hot water only. Even further control is desirable. Ideally there should be hot and cold water stop valve immediately beneath every fixture, or one in every fixture branch. Then you can re-

pair a leaky faucet without shutting off water to any other faucet.

When every fixture has been plumbed in, the long-awaited day comes when you can turn on the water. You'll get a thrill out of watching it flow from the faucet and disappear down the drain. Your plumbing system will be working.